PATTERN RECOGNITION APPROACH TO DATA INTERPRETATION

PATTERN RECOGNITION APPROACH TO DATA INTERPRETATION

Diane D. Wolff
University of Arizona
Tucson, Arizona

and
Michael L. Parsons
Department of Chemistry
Arizona State University
Tempe, Arizona

Plenum Press • New York and London

Library of Congress Cataloging in Publication Data

Wolff, Diane D., 1953–
   Pattern recognition approach to data interpretation.

   Bibliography: p.
   Includes index.
   1. Statistics—Data processing. 2. Experimental design—Data processing. I. Parsons, M. L. (Michael Loewen), 1940–   . II. Title. III. Title: Data interpretation.
QA276.4.W64 1983   519.5′028′54   83-9624
ISBN 0-306-41302-7

© 1983 Plenum Press, New York
A Division of Plenum Publishing Corporation
233 Spring Street, New York, N.Y. 10013

Printed in the United States of America

This book is sincerely dedicated to
Paul, Doug, and Ginger,
without whom the finished product
would never have become a reality.

PREFACE

An attempt is made in this book to give scientists a detailed working knowledge of the powerful mathematical tools available to aid in data interpretation, especially when confronted with large data sets incorporating many parameters. A minimal amount of computer knowledge is necessary for successful applications, and we have tried conscientiously to provide this in the appropriate sections and references. Scientific data are now being produced at rates not believed possible ten years ago. A major goal in any scientific investigation should be to obtain a critical evaluation of the data generated in a set of experiments in order to extract whatever useful scientific information may be present. Very often, the large number of measurements present in the data set does not make this an easy task.

The goals of this book are thus fourfold. The first is to create a useful reference on the applications of these statistical pattern recognition methods to the sciences. The majority of our discussions center around the fields of chemistry, geology, environmental sciences, physics, and the biological and medical sciences. In Chapter IV a section is devoted to each of these fields. Since the applications of pattern recognition techniques are essentially unlimited, restricted only by the outer limitations of the human imagination, we realize that this chapter will serve more as an illustration of the almost boundless diversity of problems and approaches in each field than a comprehensive review of published applications. We feel, however, that this goal has been well addressed by much of the scientific pattern recognition literature available today, so our time spent on it is minimized.

A second goal is to introduce the scientist to the techniques currently available. In Chapter II, a brief introduction to the mathematical foundations for each technique is given and references are provided for those interested in more detailed mathematical discussions and explanations. Particular stress is given to the interpretation and applications of the pattern recognition tool. Statistical results can be quite meaningless unless the correct questions are asked, and the data handled in an appropriate manner. Each technique has its advantages and disadvantages, and these are the focus of the discussions in these chapters. A basic knowledge of statistics is assumed, and references are listed for details beyond the scope of our present treatment.

Thirdly, we feel that there exists considerable need for a book that stresses the "how-to-do-it" approach. Many statistical books suffer the disadvantage of presenting detailed mathematical explanations of techniques with no suggestions about how best to approach specific real problems, and no references to appropriate source material. In Chapter III, approaches to problems utilizing three major statistical packages, SPSS, BMDP, and ARTHUR are considered. The availability of each package is given in Appendix V and a "how-to" step-by-step explanation and comparison of each program within these packages is provided in the chapters. The purpose of these chapters is to motivate the scientist to utilize these powerful, readily available pattern recognition techniques with his or her own experimental data. The authors will feel this book successful if we manage to motivate our readers to do so with some degree of regularity.

There remains last, but first, Chapter I. Choices are always the most difficult problem with the many statistical approaches to data analysis. In this chapter we attempt to answer questions about data choices and organization, about tool choices and their applications, about the most appropriate questions the investigator might ask of the data, and finally, about the philosophies of interpreting results.

This is by no means meant to be a comprehensive mathematically oriented statistics book. We view it more as an active tool to aid in one very important step of scientific laboratory work—namely, data interpretation.

CONTENTS

II
PATTERN RECOGNITION APPROACH TO DATA ANALYSIS

III

IMPLEMENTATION

IV
NATURAL SCIENCE APPLICATIONS

I

PHILOSOPHICAL CONSIDERATIONS AND COMPUTER PACKAGES

I.1. PHILOSOPHICAL CONSIDERATIONS

INTRODUCTION

Scientific research has become an area of enormous data production. Reams of data are routinely generated as the trend towards quantification within the sciences has increased. The need for sound mathematical methods for analyzing these data is crucial. Often data are continually produced without stopping for such analyses. The result can be the production of large amounts of inferior data. Studying the mathematical patterns underlying the data can often help to determine the best next step in the analysis and to draw meaningful conclusions from the data already gathered. Moreover, such studies may reveal that better experimental designs can be devised and implemented effectively. Also, underlying properties of the data, not directly measurable, but related to the data being produced, may be studied, and predictions related to the scientific content of the data, and future data, become possible.

This book will introduce the noncomputer-oriented scientist, as well as those familiar with computer usage, to some techniques available within most scientific communities for such data analyses. Through this study, better data collection methods may be revealed, and the time invested in the creation of inferior data or in suboptimal use of good data may also be minimized.

DATA CONSIDERATIONS

To apply mathematical analyses to scientific data, one must first thoroughly understand the scientific aspect of the problem—what are the goals for the experiments and associated data collection? It is critical to approach the data analysis problem without losing the scientific content underlying the data. This points to the importance of interaction between the scientist and the computer. Underlying structures in the data are often too complex to be unraveled by the scientist working alone. Hand calculations become impossible. But the ease of access to "number crunching" by today's sophisticated computers is often the cause of excessive data collection and its associated interpretation abuse. The computer cannot tell the difference between proper and improper applications supplied by the investigator, but can only blindly follow the computational algorithms.

Such data analysis may encompass anything from a single set of computations to a complex sequence of steps, each suggesting further procedures. The results of one step may point towards other useful analyses, or single out the ones already taken that may be unnecessary. But a complete and informed interaction between the scientist and the computer is necessary for optimal results.

The mathematical tools for these data analyses are not new. The fundamental ideas are related to the multivariate *normal distributions* derived in the 1930s. But the computer technology has had to catch up with the scientists' abilities, and it is only very recently that the two have become a compatible pair.

PATTERN RECOGNITION APPROACH

This data analysis problem is often referred to as a *pattern recognition* approach. It is the goal of such a study to identify patterns in the experimental data produced in an investigation, and to draw intelligent conclusions from them. For example, let us assume that the scientist is interested in environmental problems and more specifically in environmental air pollution. Assume that this project concentrates on the determination of the causes for the changes in visibility in a large metropolitan area from day to day. The scientist must first determine what data to consider. Without a prior knowledge of the exact causes, the scientist should generally still be able to use his or her scientific expertise to determine probable causes of this visibility variation. To begin with, broad general causes may be considered. The experimenter might start thinking about what general parameters could be important. For instance, weather, particulate concentration, or gas concentrations in the area might be considered.

To expand on these, he or she must then consider more specific knowledge of each. The weather almost certainly plays an important role in visibility. But is it the humidity, solar radiation, precipitation, wind speed, none of these, or a complex combination of all of these (and possibly others) that affect it? Are total organics, trace metal concentrations, sulfides, chlorides, or a complex combination of these important in the pollution itself? Without undertaking critical data analyses and attempting to recognize underlying patterns at each step of the study, much data may be collected that has some scientific importance but is useless in answering the scientist's original question.

In the computer-oriented pattern recognition approach each type of measurement made is considered a variable. In the above example, humidity, solar radiation, and each such measurement is a variable. Mathematically, each variable (or type of measurement) can be considered a dimension in space. For example, if at given times of the day the humidity and visibility were measured, we would consider it a two-dimensional analysis problem, and plots in two-dimensional space with values of humidity versus visibility could be made. Most problems involve many more variables than this, and therefore become problems in multivariable analyses. This is where the computer becomes an invaluable tool. The scientist has trouble visualizing more than a two- or three-dimensional space, but statistical computations are just as valid in n-dimensional space. Mathematical algorithms determine how n-dimensional calculations will be performed, but it is the job of the scientist to find the meaning of such calculations. For instance, the scientist can measure visibility, humidity, and solar radiation at given times on given days. A plot of visibility versus solar radiation, or visibility versus humidity, can be made, but usually such simple relationships are not adequate to describe the reality of the problem. What if visibility were really related by some complex mathematical function to both humidity and solar radiation? The best the scientist could do would be to generate enough measurements of the three at given times of the day and week to try to recognize patterns in the data. Computer-implemented multivariate analysis (more appropriately called "pattern recognition") can aid the scientist in this endeavor. Definitions for pattern recognition terminology are given in Appendix I, and the terms are italicized when first used in the text.

The pattern recognition approach is founded on a few basic assumptions. The first is that there *are* underlying patterns in the data; in the present case, such a variable as visibility is somehow related to solar radiation and humidity. The scientist may not know this as a fact. The best experiment, then, is to make measurements that the scientist believes are necessary to understand the problem. At least some of these experimentally determined variables must be relevant to the problem for pattern recognition techniques to be successful. If the variables are indeed related, not only will pattern recognition techniques lead to the exposure of the underlying relationships, but it will also help reveal which variables are most critical in understanding the problem. The scientist can therefore usually determine which measurements are most important in solving the problem, possibly thereby saving time and money.

PATTERN RECOGNITION QUESTIONS

Pattern recognition can be applied to a much larger variety of problems than the example given above. The critical step in the pattern recognition analysis is often that of formulating the questions to be asked, and the success of this formulation usually determines the success of the outcome. One question may be directed towards the relationship between various measurements for predicting a specific property, such as that of visibility. A second question may be directed towards fitting the data to a given mathematical model. A third approach may consist of separating the data into various *groupings* (either defined or naturally occurring ones) with the intention of determining the underlying patterns of these groupings, what the groupings are, whether the groupings are statistically valid, or what causes the groupings. It may be possible to predict to which groups unclassified samples belong according to the measurements made. For example, assume that our environmental scientist now becomes interested in water pollution. And let us further assume that the researcher would like to characterize bodies of water as either (1) too polluted to support fish life, or (2) able to support it. Now that two groupings or *categories* have been created, the question may be asked: What determines these groupings? Again, scientific expertise is necessary to determine what specific measurements could be important. The application of pattern recognition techniques will then help him or her decide which actually are the key parameters. A second question that might be asked is whether a certain body of water is able to support life (or asked in a different way, does this body of water belong to group 1 or 2?). The investigator would then make those measurements deemed to be important, using pattern recognition techniques to determine into which group that particular body of water best fits. Alternatively, it may be desirable to check the validity of the groupings, and one might, in the course of these studies, find a group of lakes that can support certain more hearty life forms, but not other fragile types. In addition, the investigator may find regrouping necessary as the data are studied in more detail. New questions may arise and therefore new answers may be sought. Many times one question will lead to the realization that more data are necessary. And the study of such data may result in the formulation of further questions. There is no one single correct approach. Each step can lead to a variety of subsequent lines of inquiry, but the results of each should be observed and then analyzed before the next step is taken. This will keep the process in line, optimize the analytical procedures, and therefore minimize the time spent on the problem.

PATTERN RECOGNITION NOMENCLATURE

One of the problems of *multivariate* statistical *analyses* is the inconsistency in the nomenclature used by various scientists. In this book, the nomenclature generally follows the descriptions by Kowalski. A brief example of such usage follows.

An *object* or *case* is an entity for which a list of characterizing parameters are made. The set of characterizing parameters is known as a *data vector* or *pattern* for the object or case. Each parameter is called a measurement, *variable,* or *feature.* The objects can be grouped according to some defined criteria of similarity into categories, or groupings. If the categories are known, and the goal of the analysis is to find the causes of such groupings, the study is referred to as supervised analysis (or *supervised learning*). If the major goal is to determine these groupings, the process is called unsupervised analysis (or *unsupervised learning*). Whether supervised or unsupervised learning is used to characterize objects and study relationships among them, it is called the *training step* since the scientist (through use of the computer) is being "trained" to recognize either defined or natural categories. It then becomes possible to add objects to the analysis whose membership in the groups is not known. This process, called the *test step,* is used to predict their inclusion in or exclusion from the groups under study. If the scientist uses test cases whose membership is known, but not given to the computer as information, a test of the prediction power of the pattern recognition analysis would be possible.

For example, returning to the example of the water environmentalist, his or her two categories are those waters where life can survive, and where it cannot. The objects of inquiry are the various lakes. Each measurement (for example, temperature, algae content, etc.), is a variable or feature. The set of measurements for a given lake is called the data vector or pattern for that lake. To determine what causes the two groupings, techniques involving supervised learning are undertaken. To determine whether any other groups (or another totally different grouping) exists, unsupervised learning can be used. Both are methods in the training step. If additional lakes about which the life functions are unknown are then considered, their investigation is called the test step and the additional lakes are considered the test set.

Now assume that at least the general questions to be asked can be determined and that data, at least some of which are pertinent to the problem, can be collected. The next step is to determine how to make this data computer-compatible (see Appendix III). The computer does not know what kind of data it is receiving. Therefore, it must be coded into computer language in some form meaningful to the computer's statistical analysis programs.

VARIABLE CODING

Variables may take on certain levels of measurement, which are determined by the methods of data collection and the inherent precision of these methods. Four major levels of data measurement exist. The first (and probably the most often encountered in the sciences) is the continuous measurement, which assumes that any value within a certain range of values is possible for the analysis. An example would be the measurement of the percent of copper in a sample, knowing that the values must range between zero and 100%. For a set of copper analyses, a subset of this theoretical range is usually present. As with all continuous measurements, the zero value is defined and meaningful, and distances between values make sense, i.e., 20% is twice 10%. In actuality, the numbers possible for the percent copper in the sample are limited in accuracy by the precision of the experimental measurement.

The second level of measurement groups data into coarsely defined regions due to the semiquantitative or limited accuracy of the measuring technique. Each group has a unique position with respect to the other data, but specific values of distances are not known. The zero point is not usually defined. An example would be to group the percent of copper in a sample as either (a) <25%, (b) 25%–50%, (c) 50%–75%, or (d) >75%. Now the order is still important, but distances [for example, group (c) is not twice group (a)] are not explicitly defined.

A third type of measurement is binary coded data consisting of those variables where dichotomies, or only two possible values, exist. An example would be a group of samples designated simply as either containing copper or having no copper. This is often the situation with a "present versus absent" dichotomy, and it is often coded as $0 =$ absent and $1 =$ present. In this case, neither distances nor rank order apply.

The fourth type of measurement refers to unordered groups. The numbers assigned are meaningless, and are used only as names or labels. An example would be coding the colors red, blue, green, and yellow as 1, 3, 2, and 4. Distances are meaningless and mathematical properties, such as adding and subtracting, are invalid. These could be replaced by four individual binary variables (i.e., the first being red, where 0 = absent and 1 = present; the second being blue, where 0 = absent and 1 = present; etc.). This, although mathematically valid, can become quite cumbersome.

Many of the statistical tools to be used in pattern recognition are affected by the level of measurement for the data. Many utilize statistical assumptions for their bases, and the validity of these assumptions in the data determine the validity of the results obtained. Most multivariate statistical analysis tools employed in pattern recognition are *parametric* tools. This means that the distribution of the variable values is known, or can be estimated. Further, a multivariate normal distribution is often assumed (see Appendix II). Tests to study the validity of this assumption will be discussed later in the text. Continuous levels of data measurements usually cause no problems. Usually each variable can be assumed to be normally distributed, and the combination of the variables jointly is also normally distributed.

The problems usually occur when the data takes on either a grouped or binary level of measurement. Pattern recognition techniques must be applied with extreme care to these variables, and interpretations should be approached cautiously. *Nonparametric statistics* should be strongly considered in these cases. Appendix VII describes this topic. The properties of these measurement types do not follow normal mathematical rules, and must be handled with care. The computer will not automatically check the underlying statistical assumptions during the coding of the data. It can only follow the program's algorithms, regardless of the data type given, and it is therefore critical for the scientist to check these types of things early on in the statistical analysis.

CATEGORIZATION OF DATA

The categories present in a data base can also be divided into four types similar to those for variables. The first, the continuous category, is, for example, like the lake eutrophication problem. It is possible to use binary categories as in the example above, i.e., whether a certain lake can or cannot support life. But in reality, it is probably more a continuous type category, one where at one extreme we have pristine conditions, and at the other extreme, totally stagnant water that can support no life. All values in between are possible. And distances are again meaningful, since if 20% of all fish life dies per year, this is a value twice that if only 10% does. Grouped data would be possible as well (0–25% die, 25%–50%, 50%–75%, >75%). Unordered categories bearing no distance-type relationships are also possible if the similarity of lakes is defined according to which variety of fish live and which die in each.

There may exist a combination of types of variables within any of these categories in a given pattern recognition analysis. Again, it is important to realize the statistical validity of the assumptions of each.

CONSIDERATIONS TO KEEP IN MIND

The choice of data to include in any study belongs to the researcher. Hopefully, he or she will have some insight into which features are sufficiently important to include. As the data analysis problem proceeds, and he or she obtains more insight into the results of prediction and the data structures discovered, new ideas may develop. But results can only be as good as the original data used, so that the choice of data and their proper coding are of the utmost importance in obtaining an intelligent solution to any problem. A common sense approach must be taken and an open mind maintained.

The meaningfulness of the results must also be determined by the scientist. The statistical methods of pattern recognition can give an indication of how well the analyses have proceeded, but they cannot create physically meaningful results. If the latter are found, they must have been present in the data, not in the statistical tools. The investigator must therefore develop good skills in the interpretation of results, while avoiding the often too attractive mistake of overinterpretations.

PROGRAMS TO BE DISCUSSED

The programs recommended and included in this book were carefully chosen to be both complimentary to each other and readily available to the scientific community. Major usage will be made of BMDP, the Biomedical Discrimination Programs from the UCLA Health Science Computing facilities; of SPSS, the Statistical Package for the Social Sciences available from SPSS, Inc.; and of ARTHUR, written by the Laboratory for Chemometrics, headed by Bruce Kowalski, University of Washington and currently available from Infometrix, Inc., in Seattle, Washington. Appendix V lists the availability of each (with addresses). Also given are references to the accompanying user's manuals and update communications. BMDP and SPSS are available at most large academic and industrial computing facilities. The third, ARTHUR, although developed primarily for the chemical and biological sciences, is available by request and is readily applicable to other fields. A less widely available, yet extremely useful complimentary package, CLUSTAN, will also play a minor part in the discussion (see Step II.4d). Section I.6 will briefly describe another program, SAS, which might be useful for IBM computer users.

We realize that our program choices are biased. SPSS and BMDP are heavily supported packages, making their applications somewhat easier than ARTHUR to implement. They also come, however, with a large yearly license fee, requiring them to be limited mostly to large institutions. ARTHUR is truly unique in the fact that it was written for the sciences by scientists. The program development has gotten ahead of the support and documentation, causing it to be harder to implement. This situation should be vastly improved with the distributions being handled by Infometrix, Inc. ARTHUR is much cheaper, with a one-time start-up fee of less than $1000, part of which goes to support the Chemometrics Society. ARTHUR is highly suggested for all scientists, with

BMDP and/or SPSS being used to complement it. Similar statistical algorithms are available in other computer statistical packages, and can be readily substituted.

POSSIBLE APPROACHES

It will be seen that the approaches and divisions detailed in Chapter II are quite variable. We have shown what has worked best in our particular applications. The serious investigator will consider this as only one possible approach, and will vary it according to the individual needs of his or her research. Some programs easily lend themselves to inclusion at several steps of the data analysis process. Often, the appropriate placement of these programs is dictated by the questions to be answered, and adjustments can be made in them accordingly.

Throughout the remainder of this book we will use full-size capital letters to designate the computer package under discussion. Words that are set in small capital letters are used to designate the individual statistical tools within each program. The first time a term is encountered that is defined in Appendix I, it is italicized. Although the techniques in BMDP are programs and in SPSS and ARTHUR are subroutines within the main program, the term "program" will be used in this book to refer to either type. Use Appendix IV, an index to the packages SPSS, BMDP, and ARTHUR, to find the given names for each statistical tool within each statistical package. The word "package" will be reserved to refer to either BMDP, SPSS, ARTHUR, CLUSTAN, or SAS.

I.2. BIOMEDICAL COMPUTER PROGRAM (BMDP)

INTRODUCTION

The first Biomedical Computer Program (then designated as BMD) manual was finished in 1961 at the University of California at Los Angeles. Continual updatings and changes have occurred over the intervening years. In 1968 an English-based control language was adopted for ease of use by noncomputer-oriented people. New program development continued, and in 1977 the BMDP manual contained 33 programs. The previous edition of BMDP, the 1975 version, contained seven fewer programs. New tools and options on statistics and graphic displays were added. In 1979, three additional programs were added, and are included in the 1979 printing of the *BMDP-77 Biomedical Computer Programs, P-Series* manual.

PROGRAM GROUPS

The series is divided into seven groups of programs. The D-series, consisting of programs designated as BMDP1D to BMDP9D, contain primarily data description programs. Each is arranged as an individual program, and although the control instructions for each program have been kept similar, each program must be accessed through a separate computer implementation. The F-series consists of BMDP1F to BMDP3F and basically produces frequency tables for the variables. The L-series to date contains only a single program, BMDP1L, which gives a life table and survival functions. The M-series deals with relationships in many of the variables simultaneously (therefore, it is called the multivariate analysis group), and groupings and variable relationships in n dimensions are included. Programs include BMDP1M through BMDP7M, BMDPAM, and the 1979 BMDPKM program. The fifth series, the R-series, contains nine programs (BMDP1R to BMDP8R plus BMDPAR) used to study various regression problems from linear to nonlinear, principal component analysis, partial correlations, and subsets of the regression problem. In 1979, BMDPLR was added, which performs stepwise logistic regressions between a binary dependent variable and a set of independent variables. The S-series programs contain BMDP1S (multipass transformations for altering the data) and BMDP3S (nonparametric statistics). The V-programs, consisting of BMDP1V to BMDP3V and BMDP8V, added in 1979, do variance and covariance analyses. Appendix IV provides the descriptions as given in the index to the BMDP manual. An asterisk is given for programs added in the 1977 edition and two asterisks for the 1979 additions. This manual, available from the University of California Press in Berkeley, California (copyright 1979), is necessary for the user of this program package. Many details and descriptions of the use of the programs are given, which we greatly encourage you to study. Definitions of nomenclature used in BMDP are also included. It is more our goal to compare what is available here with the other major programs under consideration in this book. Throughout Chapter II we will focus on such comparisons and our biases for certain applications.

Some of the BMD series programs have remained and complement those included in BMDP programs. Spectral analyses and time series are available. BMD still computes in fixed computer format (which means that the spacing of computer input is critical). Various time series programs are currently under development to be included in BMDP packages in the future.

File manipulation and data transformations can be a major undertaking in computer-based statistical analyses. In Chapter II such topics will be briefly discussed, but the major focus will be towards the simplest methods of computer data input/output to minimize problems for the noncomputer-oriented scientist.

BMDP CONTROL LANGUAGE

The control language in BMDP is English-based. Each important piece of control information is written as an English sentence using words such as "is," "are," or " = " (which all have identical meanings in BMDP). These sentences are used for commands or for making assignments. For instance, if we are considering a six-dimensional space of six measurements on each object, a sentence such as "Variables = 6." would be necessary. Note that each sentence ends with a period, as is the common practice of the English language. Sentences are grouped into paragraphs. Individual paragraphs are used for such things as describing how the data are input, what the variables are and how they are defined, and descriptions of the problem to be studied. The *BMDP Biomedical Computer Programs* manual describes in detail what paragraphs are necessary, and how sentences in each are to be written. Many options are available for each program.

Chapter III of this book will give detailed examples of computer runs using these programs. This will give the user a feeling for what actual input and output should be expected.

Users of BMDP programs should keep up with changes and updates made to the package. This is possible through the *BMD Communications*, and periodic technical reports (see Appendix V for information).

I.3. STATISTICAL PACKAGE FOR THE SOCIAL SCIENCES (SPSS)

INTRODUCTION

SPSS, the *Statistical Package for the Social Sciences,* is a system of computer programs the beginnings of which date back to 1965 at Stanford University at the Institute of Political Studies. This system of individual programs continued to develop, and in 1970 the first edition of a comprehensive new compatible program manual was printed. The development has succeeded through cooperation of social science researchers, computer scientists, and statisticians. New procedures and algorithms were incorporated. The SPSS project, which was located at the National Opinion Research Center at the University of Chicago, is now handled by SPSS, Inc. As with BMDP, a comprehensive, very useful, and instructive manual exists describing the implementation of SPSS. *SPSS, Statistical Package for the Social Sciences,* 2nd edition, published in 1975 by McGraw-Hill, Inc., with Norman H. Nie as the first author, describes Release 6 for the program. *SPSS Update,* copyrighted in 1979 by McGraw-Hill, lists changes made in Releases 7 and 8.

PACKAGE STRUCTURE

In SPSS, unlike BMDP, each statistical manipulation is a subroutine in the total SPSS package. Therefore, more than one can be accessed within a single computer runstream, and it is necessary to specify the control language only once. This gives the ability to apply multiple tools within a single run, and has the advantage of allowing more statistical types of output within a single run, but it can also tempt scientists to try to output as many statistics as possible without following an ordered sequence of steps. Therefore, we feel that although this feature may minimize the number of necessary computer runs, the advantage is only minor since in optimal statistical studies, the scientist must critically analyze each step of the process before deciding how to proceed to the next.

SPSS PROGRAMS

SPSS is arranged in 17 individual subroutines, each reflecting a varied aspect of the statistical analysis. The names of each reflect the tool to be utilized and are not grouped into series as with the BMDP programs. A list of available subroutines is given in Appendix IV. CONDESCRIPTIVE and FREQUENCIES are used for introductory data analyses. Other studies on single variables or combinations of the variables can be made through AGGREGATE, CROSSTABS, BREAKDOWN, and T-TEST. Correlations and bivariate relationships can be studied using PEARSON CORR, SCATTERGRAM, and PARTIAL CORR. Various regression models can be incorporated in REGRESSION. Variances and covariance statistics are given in ANOVA and ONEWAY. Further studies of variable relationships are found in DISCRIMINANT, FACTOR, CANCORR. The techniques, algorithms, and options for each are very complementary to those in BMDP. In this book, these comparisons of package techniques will be stressed. In some statistical problems, one package is chosen over the other owing to the ease of input, to the simplicity of interpreting the output, or to the options available. Individual applications usually determine the choice, and we can only offer suggestions here.

SPSS CONTROL LANGUAGE

Unlike BMDP, SPSS utilizes a controlled format; therefore, correct spacing on the computer cards is a necessity (see Chapter III for a more complete discussion). Columns 1–16 are reserved as the control field (similar to the paragraph names in BMDP), while the remainder of the card is used as the specification field (similar to the sentences in BMDP). These give the parameters or arguments for the particular control card. As with paragraphs in BMDP, certain control cards are required in SPSS, such as those describing variables, data, or procedures to be undertaken. No ending punctuation is necessary. Again, many options are available in each and are outlined in the manual. Chapter III of this book shows sample SPSS runs.

I.4. ARTHUR

INTRODUCTION

ARTHUR, a system of data manipulations and statistical tools, was developed at the University of Washington, Seattle, Washington, under the direction of Professor Bruce Kowalski. The project was taken over by the Laboratory for Chemometrics there under Kowalski's direction. The algorithms included in the system are those developed in his laboratory as well as by other members of the Chemometric Society (an international society for metrics or mathematical tools applied to chemical and other scientific problems). A complete detailed manual with step-by-step procedures is not yet available, but one consisting of summaries for each tool can be obtained. The 1976 version was obtainable from Kowalski. A newer, 1977 version was distributed by Alice M. Harper at the University of Georgia. Just recently, ARTHUR distribution was taken over by Infometrix, Inc. in Seattle, Washington. The newer versions contain the core programs from the 1976 version, with new and better algorithms added to compliment these. Also, program support should increase, making ARTHUR a truly unique program for the sciences.

ARTHUR is arranged, like SPSS, by subroutines. The various tools can be accessed in a single computer run, and the output of one can be utilized for the input of the next.

ARTHUR CONTROL LANGUAGE

The variation of the control format is much more restricted with ARTHUR, and it is suggested that the original data be added in ARTHUR-compatible format first (see Appendix III). With a few additional guidelines, this data format will also be adaptable to BMDP and SPSS.

The control format follows strict rules set down in the manual. The sequence of cards is important, and cards (similar to BMDP's paragraphs) for such things as inputs, format specifications, and variable lists are necessary. Each control card utilizes a free format where a control card name is given followed by a comma and a list of options for that card. The ending punctuation in ARTHUR is "$". A sample run from ARTHUR is given in Chapter III.

ARTHUR PROGRAMS

Algorithms tend to be more concentrated towards multivariate statistical analyses, although basic introductory programs exist. SCALE is used to calculate the univariate mean, range, standard deviation, kurtosis, and skewness values for each variable. Data standardization is performed to give each variable a mean of zero and a standard deviation of one. CORREL calculates the bivariate Pearson product–moment correlation matrix for the data. VARVAR is used for two-dimensional data plotting.

Three multivariate unsupervised learning techniques are available. NLM is a nonlinear mapping technique. HIER performs a hierarchical clustering analysis of the data. TREE is used to create a minimal spanning tree for the patterns. Supervised learning techniques can be performed if category structure for the data is known. KNN used the k number of nearest neighbors to define category inclusion. PLANE and MULTI perform piecewise and multiple discrimination procedures for group classifications.

Data reduction techniques are available in KARLOV, which performs Karhunen–Loeve variable combination procedures to look at the data. SELECT is used to determine which variables are best for describing the category structure. Regression analyses are available in STEP (stepwise) and PIECE (piecewise).

In the older version (1976) of ARTHUR, the above-named algorithms were the only subroutines necessary to call to implement that procedure. In the 1977 and 1980 versions, some of these have been broken up into smaller pieces so that more than one subroutine must be used to obtain the same results. This increases the flexibility in the user's choices. For instance, KARLOV (1976) has been broken up into KAPRIN and KATRAN in the new version. These changes are explained in the new manual.

One capability that has been added in the newer version of ARTHUR allows an error number to be associated with each measurement on the input data file. Patterns are then no longer considered points in space, but become fuzzy regions. A number of the procedures take these into consideration when doing the calculations and plots. This is a very new development in the program, and discussion of it will be minimized here. It is, however, a very interesting potential for the program.

I.5. CLUSTAN

INTRODUCTION

CLUSTAN was developed at the University of St. Andrews in the 1960s for studies of ecology and taxonomy. The demand grew for an access to such a program, and the first user's manual was written in 1969 to be used with the program CLUSTAN 1A. An update of the manual and program designated as CLUSTAN 1B was released in 1974. This was designed as a set of subroutine clustering techniques within a single program. The latest available edition, CLUSTAN 1C, was produced for release in 1978 at University College London. Two new analysis procedures were included along with extended explanation and examples in the manual, and the ability to use SPSS System files was incorporated. CLUSTAN was developed for the collective study of different cluster analysis methods. The input into the program can be continuous, grouped, or binary data. It will (as will SPSS and BMDP) accept similarities or distance matrices as input.

CLUSTAN FORMAT

Input into CLUSTAN follows a fixed format. There are three file specification cards, followed by the data cards, and then the procedure cards. Each procedure needs two cards: a name card and a parameter card that state the desired options to be used. These must be designed by following a fixed format. Information will be given about the programs available in CLUSTAN in Step II.4d.

I.6. SAS

INTRODUCTION

SAS, the Statistical Analysis System, was developed by SAS Institute, Inc., in Cary, North Carolina, beginning in 1966. The goal of the package development was to give data analysts a single system for all of their computing needs so as to be able to concentrate on result interpretation instead of the mechanics of implementation. The package was developed for (and is presently only available on) the IBM 360/370 computer systems. The latest manual for the package is the *SAS User's Guide, 1979 Edition* published by the SAS Institute, Inc. An abbreviated version, *SAS Introductory Guide,* is also available. See Appendix V for details.

SAS contains both simple descriptive statistical programs and multivariate analysis techniques. Each is a subroutine that can be called into the program, as needed. Therefore, only one set of data description cards is necessary to do multiple analyses. The subroutines are addressed through a "PROC" statement that lists the letters "PROC" followed by the procedure name for the routine to be used. The ending punctuation for SAS is a semicolon. No rigid column format is necessary.

SAS PROGRAMMING

SAS leaves a large amount of flexibility in programming for the user. Statements such as "IF-THEN/ELSE," "GO TO," "RETURN," and "DO" are available if the programmer would like to skip statements for certain observations or to change the sequence of execution. This can be very advantageous especially when working with large data sets. An option is available ("OPTIONS OBS = 10;") for checking errors by running the job utilizing only the first ten observations. The user can check the validity of these results before processing the total data set.

PROCEDURES AVAILABLE

ANOVA calculates an analysis of variance using balanced data. NPAR1WAY can perform a one-way analysis of variance for nonparametric data (see Appendix VII). Basic statistics are available with CORR (Pearson product–moment correlation coefficients for parametric data and Spearman and Kendall correlations for nonparametric). Univariate descriptive statistics are listed for each variable. CHART is used to produce bar charts, histograms, and pie charts. UNIVARIATE and MEANS will list the basic univariate descriptive statistics. PLOT is used for all printer plotting of results. Standard scores (see Appendix IX) are calculated with STANDARD. TTEST is used to perform and study *t*-tests.

A variety of regression analyses are available including GLM (simple regression, multiple regression, weighted and polymeric regressions, partial correlations, and multivariate analyses of variance), NLIN (nonlinear regression by least squares or weighted least squares), STEPWISE (with five stepwise regressions available, varying in the choice of criteria for adding or removing variables), and RSQUARE (all possible subset regression).

Time series analyses are available through AUTOREG and SPECTRA (multivariate). Cross tabulation tables and one-way to *n*-way frequency tables are requestable with FREQ.

Supervised learning programs include DISCRIM (discriminant analysis) and NEIGH-

BOR (nearest-neighbor discriminant analysis). Unsupervised cluster analysis can be studied with CLUSTER. Canonical correlations are present in CANCORR.

SAS can utilize BMDP programs with BMDP (which can do BMDP programs on your SAS data set). CONVERT will convert SPSS or BMDP system files to SAS data sets.

Factor analyses with choices of type and rotation are performed with FACTOR. Other programs are also available, mostly for data manipulations.

II

PATTERN RECOGNITION APPROACH TO DATA ANALYSIS

INTRODUCTION

Specific techniques from the three computer packages—SPSS, BMDP, and ARTHUR—will be discussed in this chapter. A single, comprehensive problem will be tackled, with the goal being a thorough, statistical analysis of the data base. A reasonably complex example was chosen, to illustrate both the results as well as the problems encountered in a pattern recognition study. The statistical evaluation of the data base used for the examples was performed in the authors' laboratory. Studies were made on the Maricopa County Arizona Health Department air pollution data base. Samples were taken at six-day intervals from January 1975 through December of 1977. Four monitoring stations were used, designated as MC (Maricopa County), MS (Mesa), NS (North Scottsdale), and SC (Scottsdale). All four are located within the Metropolitan Phoenix area with NS representing the most "pristine" area from a pollution standpoint. Only data from days where measurements at all four stations were collected are included in the study. Variables (and their abbreviations to be used) in the study were 24 high volume collection samples analyzed for manganese (MN), copper (CU), lead (PB), total particulates (PART), total organics (ORG), sulfates (SUL), nitrates (NIT), and chlorides (CL). Carbon monoxide (CO) data were also collected. The above abbreviations were used in the programs and will be utilized throughout the text.

Additional weather data were obtained from the Sky Harbor International Airport weather reports. These are mainly used in Steps II.3e and II.3f. A description of the pollution data base and the raw data for the pollution variables are given in Appendix III along with the methods of analyses. References to these data will be made throughout the book; therefore the reader should become thoroughly familiar with the data base.

Seven major steps are included in our approach to the data evaluation. They are numbered II.1 through II.7, the numbers corresponding to the section of this chapter devoted to the discussion of each. A summary is given in Appendix VI.

Step II.1 is a description of the preliminary data evaluations. Screenings of variables are made, compatibility is checked, and each variable is studied with respect to

17

its distribution characteristics. Step II.2 is an investigation into the stratification of the variables and natural groupings for each. Checks are also made regarding statistical differences in the groups obtained. Step II.3 is a study of sets and pairs of variables for intervariable relationships. Correlations and regressions are included. Step II.4 is the -unsupervised learning step in which all variables are simultaneously analyzed for natural groupings in the data base. This step is supplemented by the CLUSTAN package. Step II.5 contains supervised learning techniques. Two approaches can be used. The first is to check the validity of the "natural" groupings found in Step II.4. The second is to train the computer to recognize category definitions that the experimentalist knows for the data, and to study what variables affect this grouping (i.e., how do the four monitoring stations differ in their pollution readings?). Step II.6 is a discussion of the problems inherent in data reduction techniques and how different subgroups of variables in the study define the categories. The seventh step, II.7, presents the problems of data modification procedures.

It is imperative to realize that an infinite number of approaches are possible to this or any problem. The methods of choice and the order of steps used vary widely with problem types; therefore, one absolute approach cannot be obtained. Suggestions and philosophies for possible approaches, however, will be given.

Each technique will be discussed according to what it can (and cannot) accomplish, and what programs are available. A comparison of the programs and a brief mathematical background for each will be given. More detailed mathematical discussions are given in the cited references (see Appendix I).

II.1. PRELIMINARY DATA EXAMINATION

INTRODUCTION

In Step II.1, the data are studied and screened. Distributional characteristics for the variables are analyzed. Problems with missing values are considered. Erroneous entries such as keypunching errors must be found as well as identification of the valid, extreme outlying cases. Gross errors in observations, data coding, and inappropriate measurements are sought. The variable characteristics are analyzed for symmetry and normal distribution validity. Questions related to transformations and data trimming must be answered. Preliminary examination of the data to test various additional approaches are made.

STEP II.1a. COMPUTER COMPATIBILITY

The first goal of this step is to get the data into a computer compatible form (Step II.1a). A good approach to this problem is through the application of BMDP4D. This program will read any data and needs no format specifications (it assumes 80A1, meaning that each computer column is a separate entity). Column-by-column summaries of the data are tabulated. Data listings can be requested in several forms. The first possibility is to list the data exactly as they were given. The second form will replace certain key values (user-specified) with other symbols. For example, all numbers can be replaced with blanks, or letters with asterisks, and then listed. This aids in quickly identifying those places where format errors are present.

For instance, an example of the first eight patterns for the air pollution data is given in Table II.1a, Part A. See Appendix III for a detailed explanation of the data

TABLE II.1a

Part A Data Format

```
12345678901234567890123456789012345678901234567890123456789012345678901234567890

MC75JR12  1.0.04 2.96 0 1.35   2.06   79.66  11.00  2.55   0.83   1.39 1814.
MC75JR18  1.0.09 3.11 1 2.63   5.61  169.07  18.10  3.06   4.39   9.33 7303.
MC75JR24  1.0.12 1.68 1 3.79   4.84  197.43  14.84  6.37   4.31   8.07 7351.
MC75JR30  1.0.06 2.05 1 1.83   0.92   93.78   2.29  4.18   1.96   4.01 1289.
MC75FR05  1.0.31 2.01 0 1.57   2.36   86.91  10.47  2.73   4.03   3.19 3675.
MC75FB11  1.0.06 1.88 0 2.55   3.73  131.19  12.21  99999994.90   999994487.
MC75FB23  1.0.04 0.96 0 1.63   3.07   97.02  10.86  2.58   2.38   5.38 2816.
MC75MR01  1.0.13 1.12 0 3.37   4.12  204.84  14.03  7.49   4.00   9.56 3580.
```

For complete listing of the data, see Appendix III.

Part B Results from BMDP4D

FREQUENCY COUNT OF CHARACTERS PER VARIABLE

CHAR	CARD CODE	V A R I A B L E S												
		1	2	3	4	5	6	7	8	9	10	11	12	13
	BLANK	456	0	0	0	0	0	0	0	0	456	456	0	0
0	0	0	0	0	0	0	0	0	136	48	0	0	0	0
1	1	0	0	0	0	0	0	0	144	52	0	0	114	0
2	2	0	0	0	0	0	0	0	152	28	0	0	114	0
3	3	0	0	0	0	0	0	0	24	56	0	0	114	0
4	4	0	0	0	0	0	0	0	0	56	0	0	114	0
5	5	0	0	0	0	160	0	0	0	36	0	0	0	0
6	6	0	0	0	0	156	0	0	0	40	0	0	0	0
7	7	0	0	0	456	140	0	0	0	52	0	0	0	0
8	8	0	0	0	0	0	0	0	0	44	0	0	0	0
9	9	0	0	0	0	0	0	0	0	0	0	0	0	0
A	12-1	0	0	0	0	0	76	0	0	0	0	0	0	0
B	12-2	0	0	0	0	0	0	32	0	0	0	0	0	0
C	12-3	0	0	228	0	0	0	68	0	0	0	0	0	0
D	12-4	0	0	0	0	0	28	0	0	0	0	0	0	0
F	12-6	0	0	0	0	0	32	0	0	0	0	0	0	0
G	12-7	0	0	0	0	0	0	52	0	0	0	0	0	0
J	11-1	0	0	0	0	0	96	0	0	0	0	0	0	0
	similar results are listed for all alphabetics													
	12-3-8	0	0	0	0	0	0	0	0	0	0	0	0	456
TOTALS	---													
	NUMERICS	0	0	0	456	456	0	0	456	456	0	0	456	0
	ALPHABETIC	0	456	456	0	0	456	456	0	0	0	0	0	0
	SPECIAL	0	0	0	0	0	0	0	0	0	0	0	0	456

base. For this discussion, it is sufficient to realize that MC75JR12 is a code for Maricopa County, 1975, January 12. The next value listed, (1.0), in columns 12 through 14, is a code for the category (or sampling site in this case) for that set of data. A value of 1.0 is used to signify that this sample belongs to the Maricopa County sampling site. The remaining data are numerical values for the variable measurements for the sample. For samples containing missing values, a value of 9 is used in each column of that variable's format. For instance, the seventh measurement on MC75FB11 was not available and is keyed as such. See Appendix III for details of the format.

Note columns 2, 3, 6, and 7 are the only ones that should contain letters. Columns 1, 10, and 11 should contain only blanks. Column 12 should have all 1's for MC (Maricopa County sampling site), 2's for MS (Mesa), 3's for NS (North Scottsdale), and 4's for SC (Scottsdale). Column 13 should contain all decimal points. Table II.1a Part B lists a partial printout from BMDP4D for the data set. For illustrative purposes, only columns 1–13 (listed as variables 1–13) are given across the top. A similar printout would be listed for each of the eighty columns of the data. The characters for each column are tabulated vertically and totals given. Tabulations are made for BLANKS, numerics 0 through 9 (computer coded as 0–9), alphabetics A–Y (computer coded as 12-1, 12-2, etc.), and periods (coded as 12-3-8). These codes are only for the computer's usage. The numerics, alphabetics, and special characters are then totaled at the bottom.

Checks on the format for this data can be quickly made for simple gross errors. Most of the keypunching errors can be identified. Note that in Table II.1a Part B columns 2, 3, 6, and 7 contain only letters and also the blanks and decimals are appropriately placed in columns 1, 10, 11, and 13.

If, however, a character has been accidently punched in column 10 instead of all blanks as the format dictated, the number under column 10 for BLANK would be 455. Also, at the bottom for TOTALS under column 10, would be a "1" for either NUMERIC, ALPHABETIC, or SPECIAL. By using this subroutine to list the data, the mispunched card can be easily found. Special types of data listing, such as replacing key values (user-specified) with other symbols, also aids in quick identification of format errors. Realize that in the data format used in this example (see Table II.1a Part A) letters should not appear anywhere after column seven. If the data are listed with all letters being replaced with asterisks, a quick check of the listed data for asterisks will show points of data mispunching. Other replacements can also be used. The advantage of this lies in the ability to check large amounts of data much faster and much easier.

STEP II.1b. ELIMINATING FORMAT ERRORS

The next important step is to define a readable format specification for the data. Many times either errors are made in counting columns when determining the format, or some errors are not detected in Step II.1a. This step, Step II.1b, is used to make the data format compatible. Each of the three packages has the ability to find format errors. For example, a typing error in column 46 for a piece of data originally included in this

TABLE II.1b
Format Error Messages

---

1) SPSS

    *** CONVERSION ERROR IN FOLLOWING RECORD ***
    MC76JR19 1.0.10 1.58 1 2.60   3.68   173.72.12.14 7.35   1.00   13.9 6923.
                                      *

2) BMDP

    THE INTERPRETATION OF MEANINGLESS INPUT WAS ATTEMPTED.
    THE FOLLOWING RECORD IS ERRONEOUS OR DOES NOT CORRESPOND TO FORMAT SPECIFICATIONS:
    MC76JR19 1.0.10 1.58 1 2.60   3.68   173.72.12.14 7.35   1.00   13.9 6923.

3) ARTHUR
    No diagnostic information is given. A graphic beer mug is printed, and no error
    location or line information is listed. The rest of the data is ignored, and
    the program ceases to run.

---

data base was present, making this piece of data format incompatible. It was not noticed in Step II.1a. The data (with this error) were used in the three major packages. Table II.1b lists the output error message for each of the three packages. In SPSS, the incompatible lines are listed but *the program continues.* In those misformatted lines, zeros are input for variables that can not be read using the format given. The rest of the variables for that case are read normally. In this example, one variable in one case out of a total of 456 was given a wrong value of zero, but no significant calculation changes occurred, and useful data were still obtainable. Corrections could be easily made following the run. BMDP lists only the first error it encounters and does no further calculations. The disadvantage of this approach is that multiple errors require multiple runs since only the first error encountered is listed each time. Also, no further calculations are made so that the run becomes a complete loss (except for finding a single format error). ARTHUR, for this step, is not recommended. Not only is the run interrupted by a single format error present, but no indication of the format error location is given. This necessitates manual checking of the data and format statement to find the error. This often requires spending much more time than necessary.

It is then suggested to use SPSS for this step. CONDESCRIPTIVE, the introductory statistics subroutine for SPSS (described in Step II.1g), would be a good choice. If only a few cases are incorrectly formatted, the results obtained should still be representative of the total data. If the error statement indicates that all lines contain format errors, it suggests that the format statement itself is in error. If, however, only a few lines are listed, these can be checked and corrected to the proper format.

In CONDESCRIPTIVE, ranges for each variable are given, which can be used to determine whether missing values were correctly coded. If the program is reading a 999999 value as a "true" value instead of as an indication of a missing value, both the range and mean for that variable will be much too large. Therefore, after completion of Steps II.1a and II.1b, format errors should be eliminated.

TABLE II.1c
Missing Value Designation Miscoded Output from BMDP1D

CASE NO.	2 MN	3 CU	4 FE	5 PB	6 PART	7 ORG	8 SUL	9 NIT	10 CL	11 C
1	.0400	2.9600	1.3500	2.0600	79.6600	11.0000	2.5500	.8300	1.3900	1814.
2	.0400	.0300	1.5400	1.0900	76.1100	2.0300	1.9700	.5800	1.2000	099999
3	.0400	.0500	1.1800	.3800	45.7800	.3600	.3300	.4800	.1300	099999
71	.9200	.3800	34.4200	.4600	1082.7000	.6400	MISSING	1.7600	9.7500	099999

Note that 0999999.00 was written where missing values occurred for the variable CO.

STEP II.1c. DATA LISTING

If a further check on the coding of missing values is desirable, BMDP1D can be used as the next step, Step II.1c. This program will return a listing of the data similar to that given in Table II.1c. Only cases 1, 2, 3, and 71 are listed to illustrate the missing value specification. Note that in the second case (No. 2) the eleventh variable, CO, was not formatted correctly for missing values. Therefore, 999999 was read as a real number with the value of 0999999.00. After changing this to the correct code for the missing values, the word "MISSING" was written in the list in the place of 0999999.00.

The methods for coding "missing values" are given in Appendix VIII. It is realized that in most data bases of any size, values for some variables may not be obtainable for a given pattern, and ways to either manipulate these cases with some type of "average" value for the missing variables or to eliminate cases with values missing must be used. Case number 71 was included in Table II.1c to be referred to in a later discussion. Note that in this case, missing values for SUL were formatted correctly, but for CO were not. This is indicated by a value of 0999999.00 for CO and the word "MISSING" for SUL.

If Step II.1b showed no problems with missing value designations, Step II.1c could be eliminated.

STEP II.1d. VARIABLE DISTRIBUTIONS

A next logical step in the sequence is to check the distribution of each variable. This gives the user an indication of the validity of assuming a typical Gaussian (normal) distribution (see Appendix II for a discussion of this). Most mathematical tools utilized in this study assume this distribution. Although exact adherance can be relaxed somewhat, it is important to know how close the values actually reproduce a true Gaussian distribution. Program BMDP2D is used for this step, II.1d. Outliers and "atypical" values can also be found in this program. A printout for FE (variable #5) is shown in Table II.1d. To use this program, one must know approximate minimums, maximums, and ranges for the variables (easily obtained by using CONDESCRIPTIVE from SPSS for Step

TABLE II.1d

Statistical Distributions with BMDP2D for FE

```
VARIABLE NUMBER . . . . . .      5          MAXIMUM      34.0000000
        NAME . . . . . . .      FE          MINIMUM         .0000000
NUMBER OF DISTINCT VALUES .     13          RANGE        34.0000000
NUMBER OF VALUES COUNTED. .    432          VARIANCE      6.0032653
NUMBER OF VALUES NOT COUNTED    24          ST.DEV.       2.4501562
***VALUES ARE ROUNDED TO. .  1.0000         (Q3-Q1)/2     1.0000000

LOCATION ESTIMATES                                        ST.ERROR
        MEAN      2.9629630                  .1178832
        MEDIAN    2.0000000                  .2886753
        MODE      2.0000000

SOME NEW LOCATION ESTIMATES
        HAMPEL    2.4863524
        ANDREWS   2.5755856
        TUKEY     2.5487900

          Q   Q    S
M     MHAM  3      +
I.....EANE.
N     DMDA
      IPRN
```

```
EACH ''H''
REPRESENTS
  21.50
  COUNTS

MIN-----------------------------MAX
                          Q1=   2.0000000
                          Q3=   4.0000000
                          S-=    .5128068
                          S+=   5.4131191
                          EACH . =  .2635659

                    DIV. BY S.E.
SKEWNESS   5.7659494  48.9257026
KURTOSIS  61.6336126 261.4892731
```

VALUE	COUNT	PERCENTS CELL	CUM
0.	7	1.6	1.6
1.	80	18.5	20.1
2.	135	31.3	51.4
3.	100	23.1	74.5
4.	47	10.9	85.4
5.	20	4.6	90.0
6.	20	4.6	94.7
7.	9	2.1	96.8
8.	4	.9	97.7
9.	7	1.6	99.3
12.	1	.2	99.5
18.	1	.2	99.8
34.	1	.2	100.0

II.1b). This is necessary since you must specify an interval to use in a frequency count. For instance, with FE, CONDESCRIPTIVE gave a range from 0.00 to 34.0. BMDP2D was then given an interval of 1.0 to use. Values were then tabulated as being in the range 0–1.0, 1.0–2.0, etc. At the bottom of Table II.1d, these are given with a count (the number of FE measurements that took on a value within that range) and the percentage that this represents, both individually and cumulatively. Note that all FE values fall between 0.0 and 10.0 except for three cases which take values between 12 and 13, 18 and 19, and 34 and 35, respectively. These three must be studied further.

Interaction between user and computer is critical. You must ask the following: (a) Are these real values that represent days where FE concentrations were extremely high? (b) Were keypunch errors made (unlikely if Steps II.1a through II.1c were performed)? Or (c) were analytical errors made in the measurements? The tools discussed in Step II.1e will aid in identifying the cases where these extreme values were present. A decision must be made at this point of the analysis as to whether to retain these data in the set. Extreme points such as these can affect Gaussian assumptions and therefore the validity of some of the statistical routines.

The univariate statistics for FE are given at the top of the table. It should be noted that these values are calculated using the truncated data obtained for the frequency count. They will, therefore, differ from the true values found from CONDESCRIPTIVE and should not be used as the true sample statistics (although they should be similar). Note that the "NUMBER OF DISTINCT VALUES" is 13, which comes from the truncated frequency count. The "NUMBER OF VALUES COUNTED" represents the 432 cases in the data set that contained data for FE analyses. The other 24 were "missing" and not included in these statistics.

BMDP2D lists the minimum, maximum, range, variance, and standard deviation for the truncated data. A mean, median, and mode are also listed. An idea of how well the data adhere to a Gaussian distribution can be obtained from the visual histogram plot in the upper right corner. MIN and MAX points are marked, and a histogram distribution plotted. Note how the values are skewed towards the minimum value. This can also be seen by the "SKEWNESS" and "KURTOSIS" values. The skewness measures the shape of the distribution. A Gaussian curve will have a value of zero. (Note that a value of zero indicates a Gaussian curve but does not guarantee it.) Positive values indicate that the distribution has more values to the left of the mean value and negative skewness values indicate that more lie to the right of the mean. In this example, a skewness value of 5.77 indicates that the distribution is quite non-Gaussian, skewed to the left. The kurtosis value measures the peakedness of the distribution. A Gaussian shape is indicated by a value of zero. More positive values signify narrower and sharper peaked distributions. In Table II.1d, the value of 61.6 for FE indicates a very peaked non-Gaussian distribution.

The standard error listed in the center of the table estimates the potential degree of discrepancy between the true and sample means, calculated by use of the standard deviation value. If the skewness and kurtosis values are divided by the standard error, tests for normality are obtained. This is listed in the table (right side) as "DIV. BY

S.E." and values much greater than zero for both (48.9 and 261.5) indicates, again, that the data are not normally distributed.

The center of the table contains a visual line plot of the location estimates for the mean, median, mode, quartiles, and plus and minus one standard deviation. The width of the line is defined by MIN and MAX values. Also, three new robust estimates of location, Hampel, Andrews, and Tukey can be requested (left center of table). In these calculations, each measurement of FE is not weighed equally. Special considerations are given to outliers or extreme values in long-tailed distributions. For instance, the Hampel estimate eliminates the smallest and largest 15% of the FE values and calculates the mean value for the remaining 70%. The Andrews and Tukey similarly minimize the importance of outlying points by decreasing their weight when calculating a central value. See the BMDP manual for further details.

In conclusion, Step II.1d has shown that the FE distribution is non-Gaussian, skewed to the left. Much of this may be due to the three extremely large FE values found in the frequency count. It is a task now of the user to decide if this skewed Gaussian distribution is acceptable to use in his or her particular application. *conclusion*

STEP II.1e. IDENTIFYING UNIQUE CASES

In Step II.1e, use is made of BMDP1D again. Since this step is very inexpensive in relation to the amount of computer time necessary, it is well worth repeating. The goal this time is to identify these outlying cases. With a small data set, this can be performed by manually scanning the listed data. With large sets, however, it is far faster to perform Step II.1e. In BMDP1D, it is possible to have the data listed which contain values either above or below certain limits which can be specified for each variable. For instance, knowing the FE values obtained in Table II.1d, a maximum value for FE of 12.0 was specified in the input of BMDP1D. A typical pattern in the return listing is given in Table II.1e. Case 71 (which corresponds to NS75JN17) included two missing measurements, SUL and CO. It also included two values (FE and PART) that were greater than the value specified as "too large" in the input of BMDP1D. Therefore, one of the extremely large FE value identified by BMDP2D is in Case number 71. This method can be used to find all such extreme values, which alleviates the difficult task of scanning large data bases for these atypical values. One can then return to the data listing from the BMDP1D printout in Step II.1c to find what values these "too large" measurements actually took on. For example, from Table II.1c it is seen that this FE value shown as

TABLE II.1e
Data Listing from BMDP1D Including "TOO LARGE" Designation

SE O.	2 MN	3 CU	4 FE	5 PB	6 PART	7 ORG	8 SUL	9 NIT	10 CL	11 CO
1	.9200	.3800	TOO LARGE	.4600	TOO LARGE	.6400	MISSING	1.7600	9.7500	MISSING

"too large" was 34.42. This was also the case in Table II.1d tabulated as having a value between 34 and 35. Similar listings for "too small" can be requested.

It is left to the user to decide what to do with these values. It must be realized that with relatively small data bases, such extreme values can bias future calculations. Also, in future plots these will result in points far away from typical values. The next step, Step II.1f, can be utilized to recheck the Gaussian assumptions that would result if these were eliminated. By no means can these cases be forgotten, but an alternative is to omit them (temporarily at least) and rerun BMDP2D without them. It must then be decided if they are analytical errors in measurements or if they are valid values that must be explained. The method used to approach this answer is not to be found in the computer. This is completely dictated by the problem at hand and must be decided by the user who knows the analytical methods by which the data were obtained. For instance, with the air pollution data, factors such as weather, smelters in the area, and other pollution sources could contribute to extreme values. These days (and values) were noted, and the author decided to run Step II.1f (BMDP2D) again, eliminating the extreme values, at least temporarily. Special notations of what was eliminated were made. Answers to their causes were sought.

STEP II.1f.　VARIABLE DISTRIBUTIONS AFTER ELIMINATION

The rerunning of BMDP2D is shown for FE in Table II.1f. This should be compared with Table II.1d. It was decided to eliminate the two cases with values greater than 12.0. This can be seen since the "MAXIMUM" value was lowered from 34.0 to 12.0. This is also reflected under "NUMBER OF VALUES COUNTED" which has been reduced from 432 to 430. Note the change that occurred in the histogram distribution in the upper right corner. The histogram drawn in Table II.1f approaches a Gaussian distribution much more closely. The skewness value has decreased from 5.77 to 1.48, indicating that it is still skewed left, but to a much lesser degree. The kurtosis or peaked-ness value has decreased drastically from 61.6 to 2.83. The line plot in the center of the table also indicates that the distribution is much closer to normal. Notice the significant change that eliminating two cases from 432 has made in the mean value (reduced from 2.96 to 2.86). Gaussian assumptions and statistics should now certainly be much more valid (although it is still somewhat skewed to the left).

The critical problem of overmodification of the data must be carefully considered, and will be further discussed under Step II.7. Most air pollution measurements are known to take on more of a logarithmic Gaussian distribution. Whether this further data modification is made or not must be decided by the user. Overmodification must be carefully avoided.

STEP II.1g.　INTRODUCTORY STATISTICS

To this point, data have been screened and distribution characteristics studied. The final section of Step II.1 is related to introductory statistics for each variable, Step II.1g.

TABLE II.1f

Distributional Characteristics of Variables from BMDP2D

```
VARIABLE NUMBER . . . . . .   5
       NAME . . . . . .  FE
NUMBER OF DISTINCT VALUES .  11
NUMBER OF VALUES COUNTED. .  430
NUMBER OF VALUES NOT COUNTED  26
***VALUES ROUNDED TO. . . . 1.0000

LOCATION ESTIMATES
                              ST.ERROR
        MEAN      2.8558140   .0869005
        MEDIAN    2.0000000   .2886753
        MODE      2.0000000

MAXIMUM    12.0000000
MINIMUM      .0000000
RANGE      12.0000000
VARIANCE    3.2472271
ST.DEV.     1.8020064
(Q3-Q1)/2   1.0000000

SOME NEW LOCATION ESTIMATES
        HAMPEL    2.4863524
        ANDREWS   2.5755856
        TUKEY     2.5080000

                            DIV. BY S.E.
        SKEWNESS  1.4754682   12.4907483
        KURTOSIS  2.8290678   11.9749019

        Q1=  2.0000000
        Q2=  4.0000000
        S-=  1.0538075
        S+=  4.6578203

        EACH . =  .0930233

EACH ''H'' REPRESENTS 13.50 COUNTS
```

Histogram:

```
                                        H
                                        H
                                        H
                                      H H
                                      H H
                                      H H
                                      H H
                                      H H H
                                      H H H
                                  H H H H H H   H
  MIN---------------------------------------MAX

  S  -                   Q   S
  I                      3   +
  N ................................................
      HA M
      AN..E..
      MD A
      PR N
```

VALUE	COUNT	PERCENTS CELL	CUM	VALUE	COUNT	PERCENTS CELL	CUM	VALUE	COUNT	PERCENTS CELL	CUM	VALUE	COUNT	PERCENTS CELL	CUM
0.	7	1.6	1.6	3.	100	23.3	74.9	6.	20	4.7	95.1	9.	7	1.6	99.8
1.	80	18.6	20.2	4.	47	10.9	85.8	7.	9	2.1	97.2	12.	1	.2	100.0
2.	135	31.4	51.6	5.	20	4.7	90.5	8.	4	.9	98.1				

Note that the last two count values for FE from Table II-1d have been removed (FE = 18. and FE = 34.)

If many changes have been made in the data in Steps II.1c through II.1f, the program CONDESCRIPTIVE should be rerun for this step. If not, results from Step II.1b may be used and critically studied for this last section of Step II.1. When this point is reached, the user should be satisfied that all keypunching errors have been eliminated and the problems associated with extreme and missing values have been confronted. Distribution characteristics of each variable must be understood so that decisions concerning the variables may be intelligently made. Each of the three major packages have capacities to do these types of calculations. Each will be discussed.

In ARTHUR, the program to use is SCALE. The major disadvantage of this program, as seen by the authors, is that no simple way of considering subgroups of patterns exists. For example, in our problem it would be informative to have information not only on the data base as a whole, but also on MC, MS, NS, and SC individually to see what variations existed between them. For SCALE, this is usually accomplished by five individual computer runs, one for the data base as a whole and one each with the data base including only those patterns at each station.

In BMDP, no distinct program exists for such calculations. But from Table II.1d it can be seen that many of these values are obtainable from BMDP2D (although results from this program use the truncated values). Other BMDP programs do list some simple statistics.

SPSS CONDESCRIPTIVE has subgroup capabilities. By defining each category as a subfile in the input of these programs, each will be calculated individually if the program is told to run EACH subfile. (See Chapter III for details on this.) It can be designated to run ALL subfiles as well, so that the five desired results can be obtained within a single run. A typical printout from CONDESCRIPTIVE is given for FE in Table II.1g. This includes in Part A the results from ALL subfiles and in Part B the MC results obtained from EACH subfile. Additional printouts for the other three sampling sites would also be obtained in this part.

CONDESCRIPTIVE lists values for the mean, variance, range, standard error, kurtosis, minimum, standard deviation, skewness, and maximum. It is instructive to compare the total data set statistics (Part A) with those for individual subgroups of samples such as the sampling stations.

SPSS also has a subroutine, FREQUENCIES, which was developed for noncontinuous data. In these problems, statistics such as modes and medians have more meaning than mean values. For noncontinuous data, FREQUENCIES can also plot frequency histograms. This subroutine also has the options of "ALL" or "EACH." If FREQUENCIES is applied to continuous data, results similar to those from CONDESCRIPTIVE will be obtained.

Part C of Table II.1g lists the results from FREQUENCIES for FE in the category MC. Results similar to those obtained in Part B of the table are seen.

In BMDP, BMDP5D gives results similar to FREQUENCIES. Large histograms can be plotted and categories can be combined into groups and used. In BMDP5D, normal and half-normal probability plots can also be requested. These can be used as well as Step II.1g to determine the normality of the data and to identify outlying points. See the manuals for details.

TABLE II.1g

Univariate Statistics for FE

Part A **CONDESCRIPTIVE** for RUN SUBFILES ALL

VARIABLE FE

MEAN	2.853093	STD ERROR	.8526988-01	STD DEV	1.768194
VARIANCE	3.126509	KURTOSIS	3.409757	SKEWNESS	1.597644
RANGE	12.48000	MINIMUM	.000000	MAXIMUM	12.480000

VALID OBSERVATIONS - 430 MISSING OBSERVATIONS - 26

Part B **CONDESCRIPTIVE** for RUN SUBFILES EACH - Results for MC

VARIABLE FE

MEAN	2.643981	STD ERROR	.1389977	STD DEV	1.444506
VARIANCE	2.086598	KURTOSIS	1.2934170	SKEWNESS	1.097730
RANGE	7.920000	MINIMUM	.0000000	MAXIMUM	7.920000

VALID OBSERVATIONS - 108 MISSING OBSERVATIONS - 6

Part C **FREQUENCIES** for RUN SUBFILES EACH - Results for MC

FE

MEAN	2.644	STD ERR	.139	MEDIAN	2.305
MODE	1.570	STD DEV	1.445	VARIANCE	2.087
KURTOSIS	1.214	SKEWNESS	1.083	RANGE	7.920
MINIMUM	.000	MAXIMUM	7.920		

VALID CASES 108 MISSING CASES 6

This Step

After successful completion of Step II.1, each variable has been studied individually and screening for errors has been completed. The next major step is to study the stratification of the variables into groups. Appendix VI gives a summary of Step II.1 as well as the rest of the steps.

STANDARD SCORES

At this point in an analysis, the investigator must decide if data need to be scaled to create standard scores (or z-scores). If variables of different absolute magnitudes are used simultaneously, some type of *scaling* must be used to initially regard each feature as equally important in the training procedure. The most common method is to give each variable a mean of zero and a standard deviation equal to one. This is discussed in Appendix IX with respect to each of the three major packages used.

II.2. DATA STRATIFICATION

INTRODUCTION

section

In the last part of Step II.1, we discussed how to obtain statistics for the various groups (or categories) on an individual basis. Often one is unsure if these groups are really distinct. One method of investigating this is through the use of CONDESCRIPTIVE in SPSS (Step II.1f). However, the problem with utilizing this program is that the comparison of ranges, means, etc., is left completely to the subjectivity of the user.

In Step II.2 we will discuss statistical criteria for comparing groups. Of course, often no preconceived idea of groupings is available. In this case groups can be defined as discussed in Step II.4 by unsupervised learning before applying Step II.2. The programs discussed in Step II are used on preexisting groups to answer such questions as:

1. Are the data homogeneous or stratified into groups?
2. Considering the individual variables, are the data homogeneous or stratified into groups?
3. Are group differences statistically significant?
4. Are other outlying points now seen?
5. What are the differences in the categories?

t-TESTS

The first procedure in this step is to calculate t-test statistics for the data base. The group of samples being considered is assumed to be taken from an infinite-sized population. This means that the data we are considering for MC are hopefully representative of the overall pollution profile of MC. The sample mean ($\bar{x}$ is an estimate of the true population mean (μ). Or in other words, μ is inferred from $\bar{x}$. To compare two groups of data such as MC with MS, one can ask the question: Are these two sample means indicative of a single population? It is possible that two samples drawn from the same parent population could appear to be different due to the variability within the population. Therefore, different means do not necessarily indicate that the populations from which the two sample sets were drawn actually differed in the characteristics being studied. A test is used to determine if the samples were drawn from a single population. The hypothesis used assumes that they were. If this test fails, one can infer that the data were obtained from different populations. This test is called the Student's t-test and is available in BMDP3D or TTEST in SPSS. The goal of this step, Step II.2a, is to compare pairs of categories and their means to obtain univariate statistics and histograms within each. A t-test is then performed to compare the means of the groups, and to decide if the two groups are from the same parent population.

The Student's t-statistic assumes that the variable to be studied approximates a Gaussian, or normal distribution (see Appendix II). Theoretical values for t are calculated and tabulated which indicate the maximum differences expected between the true mean and a given sample mean for a strictly Gaussian population with a given standard deviation. These tabulated t values vary with the number of measurements in the sample and the confidence with which the test is to be made (see a basic statistics book for details of the calculation). T-values depend upon the degrees of freedom (number of samples -1), since the smaller the sample size, the less likely the sample mean will reflect the true population mean. Also, they depend on the standard deviation since a larger standard deviation indicates more variation of possible values within the population, and it is therefore less likely that the sample mean will reflect the population mean.

The t-statistic can then be used to compare the means of two sample groups at a specified confidence level to see if they were drawn from the same parent population (or possibly infer different parent populations). A pooled t-value is calculated which compares the means of the two groups and a separate t-value is calculated which compares the separate means to the parent population. A P value is then calculated which is the probability that a larger difference in the two t-values calculated above could still represent the same parent population.

A t-test usually proceeds as follows: Assume that the population means are the same with a given confidence value. Now use the t-statistic to calculate a pooled and a separate value. From the tabulated frequency distribution of this statistic find the probability of getting a more extreme value than that calculated which still has both means equal. If this value is less than the given accepted probability, reject the hypothesis and infer the means are truly different.

The 95% confidence level is often used. However, tabulations are available for other values. Also, in many cases, the goal may not be to determine if the means are of different populations, but to infer whether the first is larger (or smaller) than the second one. This is called one-tail testing. Since the t-statistic includes a term containing the difference of the means, the sign to expect with this calculation can readily be obtained.

A second way to compare two categories is to compare the variances. An F test will perform this function. This is often used when more than two categories are simultaneously considered. Details of the test will be given in Step II.2b. In this application, the variances for the two categories are arithmetically divided. A tabulation is available of the probability that certain F values occur for sample groups taken from a Gaussian parent population. The user again decides the confidence level acceptable to him, and if the calculated value exceeds the tabulated value, the equality of variances hypothesis can be rejected.

STEP II.2a. COMPARISONS OF TWO GROUPS

Both BMDP3D and TTEST in SPSS give t and F tests. However, BMDP3D will also plot histograms of the results. Means, standard deviations, standard errors of the means, the number of cases included, and the minimum and maximum values for each group individually are given. Individual group histograms are also plotted for visual comparisons of the two samples.

Table II.2a presents a sample result using BMDP3D to compare the variable FE using two-pair combinations of the four groups available: MC, MS, NS, and SC. The goal is to compare these categories pairwise to see if differences present in the means and standard deviations are really significant. It must be realized that if either a large number of variables or a large number of categories exist, the output of this subroutine can be enormous.

The first section of the table compares FE in groups MC and MS. In the center, means, standard deviations, standard errors of the means, minimums, and maximums for each category are listed. These aid in initial studies of the distributional characteristics of the two groups. On the right side, histograms for each are plotted side by side. MC, containing 108 cases, is shown as the first histogram. The symbol used is an "H," each of which represents four cases. Next to that is the histogram for MS, containing 108 cases as well, and utilizing an "X" to represent each 4.3 cases. This histogram, although rather crude, can give visual indications of the distributions that exist.

On the left the t and F statistical results are listed. The t-statistic has a value of 1.04 which corresponds to a calculated P value of 0.297. This indicates that there is a 29.7% probability that a larger difference in the means than really occurs (2.6440 versus 2.4307) could still reflect the same parent population of FE values. This is greater than the value of 5% that is usually used as the accepted value. Therefore, the hypothesis is accepted. This means that the mean values that were found for FE in MC and MS are not statistically significantly different.

The F value compares the variances. The F value obtained when comparing MC and MS was 1.16. This is associated with a probability (P) value of 0.456 from the stored F tables. This indicates that there is a 45.6% chance that a larger difference in variance could reflect the same parent population. Therefore, the variances are also assumed to be sampled from the same parent population with no statistical differences. This can be seen qualitatively from the histograms.

TABLE II.2a

Comparison by *t*-Test of Group Means and Variances for FE from BMDP3D

MARICOPA COUNTY AND MESA

VARIABLE NUMBER 5 FE

GROUP	1 MARI (N= 108)	2 MESA (N= 108)
MEAN	2.6440	2.4307
STD DEV	1.4445	1.5527
S.E.M.	.1390	.1494
NUMBER	108	108
MAXIMUM	7.9200	8.9100
MINIMUM	.0000	.4500

STATISTICS		P VALUE	D. F.
T (SEPARATE)	1.04	.297	212.9
T (POOLED)	1.04	.297	214
F(FOR VARIANCES)	1.16	.456	107,107

(handwritten: Levenes δVar (F), (F))

```
                           1 MARI                    2 MESA
                           HH                        X
                           HHH                       XXXX
                           HHH                       XXXX
                           HHHHH                      XXXXX
                           HHHHHH                     XXXXX
                           HHHHHHHHHHH H     XXXXXXXXX XXXX
                   MIN-----------------MAX MIN-------------MAX
                     AN H =   4.0 CASES    AN X =   4.3 CASES
```

MARICOPA COUNTY AND NORTH SCOTTSDALE

VARIABLE NUMBER 5 FE

GROUP	1 MARI (N= 108)	3 NSCT (N= 106)
MEAN	2.6440	3.9166
STD DEV	1.4445	2.2435
S.E.M.	.1390	.2179
NUMBER	108	106
MAXIMUM	7.9200	12.4800
MINIMUM	.0000	.3000

STATISTICS		P VALUE	D. F.
T (SEPARATE)	-4.92	.000	178.8
T (POOLED)	-4.94	.000	212
F(FOR VARIANCES)	2.41	.000	105,107

```
                           1 MARI                    3 NSCT
                           HH                         X
                           HHH                        X X
                           HHH                        XXXX
                           HHHHH              XXXXXXXX X
                           HHHHHHH H          XXXXXXXXXX X
                           HHHHHHHHHHH H      XXXXXXXXXXXXXX X
                   MIN-----------------MAX MIN-------------MAX
                     AN H =   4.0 CASES    AN X =   3.0 CASES
```

Note: similar results would be listed for the pairs: MC/SC; MS/NS; MC/SC; and SC/NS.

For the second case, FE in groups MC and NS, note first that the histograms are different. The two statistics also show that the probability of the mean value of FE for these two groups representing the same parent population is essentially zero. Therefore the means and variances in these groups are significantly different.

This procedure continues for each pair of groups. A summary table can then be prepared of results.

In a study designed to see variation and grouping of multidimensional data it is desirable for values of means in different groups to be significantly different. If this is not the case, the data for the particular variable may be providing redundant information and may not be useful in the overall evaluation of the data base. There is also the possibility that there is redundancy in the groups as defined.

ANALYSIS OF VARIANCE

The major disadvantage of the approach presented in Step II.2a is the large number of comparisons necessary to study all possible pairs of the variables present. An analysis of variance can be used to study each variable in any number of groups simultaneously. In this case, the variances of two or more groups are analyzed simultaneously. In the two group cases, this method is mathematically equivalent to that in Step II.2a, the t-test. However, if more than two groups are present, different calculations are necessary.

In an analysis of variance study, the null hypothesis to be tested states that the groups present are simply independent samples drawn from the same parent population (or different parent populations having the same means). The effect of random fluctuations in the data will normally result in some variation in the mean values. This statistical test is then used to determine if the differences present in the data are due to these random fluctuations or true differences in the parent population. This is a straightforward expansion of the null hypothesis used in the t-test.

The general rationale of an analysis of variance is the assumption that the total variance present in the data can be separated into two types. The first is the variance within the groups. This reflects the spread of the data within each group, or the variability present within the individual groups. The other type of variance is the variance between groups. This reflects the actual variation present between the groups. This may be due to the effect of some applied treatment or may just be a function of chance. If there are real differences between the groups, one would expect the variance between groups to be large relative to the variance within groups. An analysis of variance is then used to determine if this difference between groups is a function of chance or not.

In the example here, an analysis of variance will be used to decide if the means in the pollution measurements are due to random fluctuation within the same parent population for the four groups (MC, MS, NS, and SC) or due to real differences. The first option would indicate that the pollution does not vary in the different areas of the Phoenix valley.

The measure of variability used for most analysis of variance procedures is the sum of squares (SS). This refers to the total sum of the squared differences between a set of individual values and their mean. Three such values are calculated:

1. Total sum of squares—Sum of squares of the deviations of each sample from the grand mean (which is calculated as the mean of all samples taken together as a single group). This is an indication of the variance present in the complete data base.
2. Sum of squares within groups (SS_w)—Calculated as the sum of the squared deviations of each sample in each group from that group mean. This gives an indication of the variance within the groups.
3. Sum of squares between groups (SS_b)—The sum of the squared differences between each of the group means and the grand mean of the data base, weighting each deviation for the number of cases present. This indicates the variance between groups.

Estimates of the population variance are then calculated and used to test the null hypothesis. Both SS_w and SS_b are used, but must be adjusted to reflect their appropriate degrees of freedom in order to estimate the variance. These are then called the mean squares, and are designated as MS_w and MS_b, respectively. MS_w reflects the variance within the groups and MS_b, the variance between the groups.

An F-ratio is calculated to compare these two values. This is defined as

$$F = \frac{MS_b}{MS_w} = \frac{SS_b}{k-1} \frac{n-k}{SS_w}$$

where k is the number of groups and n is the number of samples. The MS_b reflects differences between the groups, and MS_w the variations within the groups. Therefore, as the differences between groups increases, the F-ratio increases as well. Like the t-test, a tabulation of values expected is available, and the result from a given problem is compared to these. If the experimental value is greater than that tabulated (for a given probability and appropriate degrees of freedom) the null hypothesis is rejected and true differences in the group populations assumed.

STEP II.2b. COMPARISONS OF ALL GROUPS SIMULTANEOUSLY

In the analysis of variance previously described, only one independent variable is considered at a time. This is known as a one-way analysis of variance. This procedure, comprising Step II.2b, can be found in ANOVA in SPSS or BMDP7D. No equivalent program is available in ARTHUR. BMDP7D will be described here. In this case, an analysis of variance is used to test the equality of the group means and to plot side-by-side histograms of each group.

Results for FE in the four groups are shown in Table II.2b. Across the top, MARI, MESA, NSCT, and SCOT indicate that everything below this column pertains to that category. Near the bottom, note that the same statistics are listed for each group individually as in Step II.2a, but that these are now listed for all four groups side by side for easier comparison. All four histograms are also plotted side by side with the mean value designated by the variable M on the histogram. At the bottom, statistics are listed for all groups combined. An analysis of variance is then performed to test the equality of the means.

The sum of squares of the difference between the sample value and the mean is calculated within each group individually and between the combined group. The total sum of squares is 1341, with 1179 of it being accounted for within the individual groups and 162 of it being between the groups. An F ratio is calculated, and a probability figure given as the probability of exceeding that F ratio when the group means are really equal. In the case of FE, there is essentially no probability of obtaining an F value of 19.54 with the different group means coming from the same parent population. Therefore, it is decided that the means represent different populations. In this case the FE pollution must vary in the Phoenix valley.

It is worthwhile to note, at this point, that although BMDP3D is a slower process with many more results, BMDP7D indicates only that the means of the four groups are not the same, but does not indicate whether this is due to all four being different or if only one or two differ (as was the case in this example). There is no indication from BMDP7D that it is NS that varies in FE with the other three groups being statistically similar. Therefore, a suggestion might be to utilize BMDP3D if a manageable number of variables and groups exist. If not, BMDP7D could be performed first, and those variables whose means were significantly different could be studied in more detail with BMDP3D.

This technique can also be used for data reduction to aid in the removal of "constant" variables. Care, however, must be exercised at this point. Choices in data alterations may be crucial in the success of the analysis. See Step II.6 for details.

ONEWAY in SPSS will give results similar to those of BMDP7D. Histograms similar to those in Table II.2b are not available. Trends across the categories can be investigated, and *a priori* and *a posteriori* contrasts made. Comparable results are obtained with both programs, although we feel BMDP7D is superior in the display of results.

For checking classification separation, a one-way analysis of variance is sufficient. For other types of statistical evaluation, more than a one-way analysis may be necessary. The basic ideas of these will be addressed in very brief detail here, since multiway analysis is rarely applicable to simple classification procedures. It is, however, still very useful in many scientific endeavors.

MULTIPLE WAY ANALYSIS OF VARIANCE

A multiple way analysis of variance is available in ANOVA in SPSS. In this case, more than one independent variable is considered and the combined effect of all of them on the classification is studied. In the normal types of classifications we have considered,

Group Studies for FE with Histograms and Analyses of Variances from BMDP7D

```
TABULATION OF VARIABLE      5 FE      WITH STRATIFICATION ON VARIABLE    2 GROUP

        MARI          MESA          NSCT          SCOT          OTHER
         .........................................................+
VAR  5
EXCLUDED
VALUES
        ******        ******        ********      ******

        TABULATIONS AND COMPUTATIONS WHICH FOLLOW EXCLUDE VALUES LISTED ABOVE

MIDPOINTS
  19.000)
  18.000)
  17.000)
  16.000)
  15.000)
  14.000)
  13.000)
  12.000)                          *
  11.000)
  10.000)
   9.000)          **              *****
   8.000)*         *               **
   7.000)          **              *******              *
   6.000)******    **              **********           ***
   5.000)******    *********       **********           ****
   4.000)***********  *********     M**************      *************
   3.000)M*********************25 **********************25 *********************25 ***********************25
   2.000)***********************41 M*********************37 ********************** M***********************38
   1.000)***********************  ********************30 *********** ***********************25
    .000)***       *               **                  *
  -1.000)

GROUP MEANS ARE DENOTED BY M'S IF THEY COINCIDE WITH *'S, N'S OTHERWISE

                                                              ALL
           MARI       MESA       NSCT       SCOT       COMBINED
MEAN       2.644      2.431      3.917      2.441       2.853
S. DEV.    1.445      1.553      2.244      1.258       1.768
N          108.       108.       106.       108.        430.
MAXIMUM    7.920      8.910     12.480      7.350      12.480
MINIMUM    .000        .450       .300       .250        .000

                    SUM OF SQUARES    DF    MEAN SQUARE   F RATIO
        BETWEEN       162.2433         3      54.0811     19.5403
        WITHIN       1179.0291       426       2.7677
        TOTAL        1341.2724       429

                                    PROB. F EXCEEDED
                                          .0000
```

we have not addressed this type of "synergistic effect" relationship. This becomes an entirely different approach to scientific statistics.

Assume a biologist is experimenting with chickens, and the categories are healthy versus unhealthy, each of which are determined by a set of size and weight requirements. Now assume that a study is undertaken to check nutritional requirements for the chickens. The variables chosen might be (a) total bulk of feed, (b) vitamin A content of the feed, and (c) zinc present in the feed. The relationship of these three variables to the defined categories must be considered. The effect is rarely an additive result of the three. More often, there are variable interactions as well. The sum of squares can still be divided between (SS_b) and within (SS_w) the groups, as with the one-way case. Now, however, interaction terms must be added as well. These will be designated as SS_{AB} for the interaction between the groups due to the interaction of variable A with variable B. Therefore, in the two-variable case (A and B) this becomes

$$SS_t = SS_w + SS_b$$
$$= (SS_A + SS_B + SS_{AB}) + SS_b$$

Now an F statistic can be derived to check the effects of variables A and B on the categories. In this example, four different F values are needed. These are used to check the statistical significance of category classifications due to

a. both variables considered as a whole;
b. the interactions of the two variables,
c. each variable considered individually.

In ANOVA in SPSS, up to five variables can be studied simultaneously. Details are given in the manual.

STEP II.2c. VARIABLE SUBGROUPS

Subgrouping of a given variable is described in Step II.2c. For instance, it may be instructive now to study whether the FE concentration varies during the different days of the week. BMDP9D can be used to handle this problem. Any number of groups can be considered and several levels for several variables can be studied. The main limitation is the manageability of the results. Table II.2c illustrates an application of this program to the study of FE variations on a daily basis from station to station.

In the table, the samples are divided into 28 subgroups (CELLS), separated by the day of the week and the sampling station. The frequency occurrence of each is given along with the mean and standard deviation value for FE in that subgroup. A mean and standard deviation for the entire data set is also calculated (see upper right corner of the table). If the variable chosen to define the subgroups is continuous, ranges can be used to create groupings in the data. For instance, if we wanted to study the FE concentrations at given MN values, then FE could be studied within given ranges of

TABLE II.2c
BMDP9D Considering Each Group Separately

CELL NUMBER	DAY 16	CAT	VARIABLE 6 FE FREQ.	MEAN	STD.DEV.	MEAN 2.8531 ST.DV. 1.768 S	S	M	S	S
1	SUN	MC	12.	2.47917	1.5086		A			
2	MON	MC	20.	2.59050	.9561		B			
3	TUES	MC	18.	3.02333	1.4684			A		
4	WED	MC	12.	2.17083	1.5034	A				
5	THUR	MC	12.	2.79083	1.9291			A		
6	FRI	MC	18.	2.96611	1.4645			A		
7	SAT	MC	16.	2.29000	1.4641	A				
8	SUN	MS	12.	2.01917	1.4158	A				
9	MON	MS	20.	2.60650	1.5433		B			
10	TUES	MS	18.	2.79222	1.5080			A		
11	WED	MS	12.	2.11000	.9961	A				
12	THUR	MS	12.	2.59333	2.1044		A			
13	FRI	MS	18.	2.55111	1.7673		A			
14	SAT	MS	16.	2.09625	1.4299	A				
15	SUN	NS	12.	3.43000	2.1142				A	
16	MON	NS	20.	3.90150	1.7669					B
17	TUES	NS	16.	4.84187	2.8088					A
18	WED	NS	12.	3.19583	1.7709			A		
19	THUR	NS	12.	4.16833	2.5828					A
20	FRI	NS	18.	4.00000	2.5790					A
21	SAT	NS	16.	3.52062	1.8989				A	
22	SUN	SC	12.	2.12667	1.2059	A				
23	MON	SC	20.	2.43650	1.1651		B			
24	TUES	SC	18.	2.96389	1.6672			A		
25	WED	SC	12.	2.05083	.9952	A				
26	THUR	SC	12.	2.61667	1.7123		A			
27	FRI	SC	18.	2.62111	1.0113		A			
28	SAT	SC	16.	2.05062	.7476	A				

EQUALITY OF CELL FREQ.	CHI-SQUARE	17.73953
	D.F.	27.
	ALPHA(CHI-SQ)	.9113
WITHIN	SUM OF SQUARES	1122.65302
	D.F.	402.
	MEAN SQUARE	2.79267
BETWEEN	SUM OF SQUARES	218.61946
	D.F.	27.
	MEAN SQUARE	8.09702
EQUALITY OF MEANS	F-VALUE	2.89938
	D.F.-S	27. 402.
	ALPHA(F)	.0000

MN. Typical ranges would be obtainable from the values used in BMDP2D in Step II.1d for the histograms.

The data in each cell are plotted in the upper right side of the table. The letter plotted represents the frequency of occurrence of that value for FE. The horizontal axis, "SSMSS," in Table II.2c represents plus and minus one and two standard deviations (the first and second "S," respectively, on each side of the mean, "M"). For frequencies above 9, the letters A through I are used for the ranges 10–19, 20–29, 30–39, etc. up to 90–99; and J through S are used for 100–199, 200–299, etc.

The data in the various cells are then compared. A chi-square test is performed as well as a one-way analysis of variance. The chi-square test is used to check the equality of the cell frequencies. The analysis of variance (F-test) studies the equality of the means. Interpretation of the F-test was discussed in the previous step.

CHI-SQUARE TEST

The chi-square (χ^2) test is really a nonparametric test. This means that the data to be analyzed need not follow Gaussian assumptions (see Appendix X). It is used to test hypotheses relating frequency counts in various categories. Cells such as those in this step are developed, and counts in each are recorded. The null hypothesis is the assumption that there is no significant difference between the observed frequency distribution and the hypothesized frequency distribution. The hypothesized model must be user-defined and may be completely theoretical. For a cell structure, it is usually defined as the frequency distribution that would be expected if the two variables were completely independent of each other. The chi-square test is then used to check if the differences present are due to random fluctuations or if they are truly significant. This statistic is calculated as

$$\chi^2 = \sum_{i=1}^{k} \frac{(O_i - E_i)^2}{E_i}$$

where O_i is the observed frequency for cell i, E_i is the expected frequency for cell i, and k is the number of categories or cells.

To perform a χ^2 test, a table of χ^2 values is necessary. The value chosen is dependent upon the degrees of freedom and an assumed probability level. If the calculated χ^2 is greater than the tabulated value, the null hypothesis is rejected. Therefore, it is concluded that a difference between the observed and expected values does indeed exist at that level of significance. If the degrees of freedom equal unity, the continuous curve from which the tabulated values are derived tends to underestimate the actual probability. A Yate's correction is made to adjust for this. Such a correction is made and stored in the computer in the χ^2 tables. To apply a chi-square test, the categories must be mutually exclusive. This means that a sample cannot belong in more than one cell simultaneously. Also, at least five samples should be represented in each cell (although some programs can handle empty cells).

With computer implementation of both the F test and the chi-square, a level of significance need not be specified. The computer has both types of tables stored for all significance levels. The result of the comparison between expected and observed values is then given as an alpha value. The alpha is defined as the probability of exceeding these F or chi-square values and still having the cell mean differences (or cell frequency distributions) represent the same parent population. The alpha value, therefore, is essentially the significance level at which the null hypothesis can be accepted.

In Table II.2c (from BMDP9D), the chi-square and F tests are performed to check the cell distributions of FE over the days. The chi-square test gives a χ^2 value of 17.7 with 27 degrees of freedom. The alpha calculated for this is 0.9113. The F test divides the sum of squares value between (SS_b) and within (SS_w) the categories. The SS_b value is 1122.7 with 402 degrees of freedom. The SS_w value is 218.6 with 27 degrees of freedom. After adjusting these values for their degrees of freedom to obtain the mean square values, the F statistic is calculated (see Step II.2b for details). This value is calculated as 2.90 which relates to an alpha value of 0.0000.

Since the alpha value is defined as the probability of exceeding these F or chi-square values and still having the null hypothesis valid for mean differences (or cell frequency distributions), it can be concluded that the mean FE values vary significantly with day or station sampling. This does not, however, indicate which are different and which are the same (if any). Further, there is a 91% probability that the frequency distributions are equivalent (chi-square test).

Many different questions could be posed with regard to the data with this program. The user must once again provide intelligent questions to derive useful information from the data.

OTHER SIMILAR PROGRAMS

SPSS BREAKDOWN will provide similar information to BMDP9D. Complex classifications for one to five independent variables may be used. An analysis of variance is given but no chi-square. A test of linearity is also available for one-way breakdowns. More variety for the output design is available including a tree-diagram style output. No histogram, however, is obtainable.

II.3. INTERVARIABLE RELATIONSHIPS

INTRODUCTION

During the first two steps, a thorough study of each variable and category distribution was undertaken. The next step in the process is to study intervariable relationships. If two variables have an *a priori* interest to the investigator, the process described in Step II.3b can be applied first to study bivariate plots of these. If no *a priori* interest in specific variables is present, Step II.3a should be performed first.

CORRELATION COEFFICIENTS

Step II.3a is used to study relationships between the variables by calculating intervariable correlations. Correlation coefficients are obtained which are summary values that indicate how strong a relationship exists between two variables. Usually the Pearson product–moment correlation coefficient is calculated, which utilizes linear relationships. For this, a value of 1 indicates a total relationship, where an increase in the value of the first variable is reflected in an increase in the second. The bivariate graph for the two variables would show a straight line having a positive slope. A value of −1 also indicates a total relationship. In this case, however, an increase in the value for the first variable is accompanied by a decrease in the second. The linear graph representing these two would then have a negative slope. As the correlation coefficient value decreases from 1 (or increases from −1) to approach zero, more scatter in the plot of the two variables away from the line would be expected. When the value reaches zero, an absence of a linear relationship between the two variables is present. Mathematical details for this can be found in a basic statistics book.

STEP II.3a. CALCULATION OF CORRELATION COEFFICIENTS

Correlation coefficients are calculated for pairs of variables. It is often useful to study the correlations for all possible pairs simultaneously if no *a priori* prejudice towards certain variable pairs is present. The results become extensive, however, if the data base contains a large number of variables. It would also be instructive to study the relationships first by categories and then for the data base as a whole for comparison. PEARSON CORR in SPSS will allow this. Again, as with CONDESCRIPTIVE (see Step II.1g), one can ask for subfiles EACH and subfiles ALL within one run so that the correlations in each of these can be easily compared. With ARTHUR, no subfile structure can be defined for the correlation program (CORREL) which necessitates five runs to obtain the same results: MC, MS, NS, and SC individually as well as the correlations for the total data base. No correlation program, as such, exists in BMDP, although individual correlations can be calculated when two-dimensional plots are made (see Step II.3b). Therefore, for Step II.3a, SPSS program PEARSON CORR is suggested. A printout from this program for ALL subfiles is shown in Table II.3a. The file called "NONAME" is used which contains the subfiles MAR(MC), MESA(MS), NSCT(NS), and SCOT(SC). Therefore, the correlations listed here are for the data set as a whole, requested by using a subfiles ALL indication.

Across the top of the table and along the left side are listed all variables in the data set. The correlations calculated for all possible pairs of variables are listed in the table. The top number in each group of three is the Pearson product–moment correlation coefficient for the two variables. Reading across the top data line, 1.000 is the value for MN with MN (expected since an increase in MN is always reflected by an increase in itself), −0.0527 is the value for MN with CU, 0.7952 for MN with FE, and so on. Note that both halves of the table are given. Circled are the correlation coefficients for MN with CU and CU with MN. These values must be identical. It is, therefore, redundant to list both as PEARSON CORR does, but no problems result because of it. Consequently only the upper right half (given in the triangle) is really necessary.

With this printout format, it can be rather hard to visualize trends, due to the

TABLE II.3a

Pearson Product Correlation Coefficients for All Data

	MN	CU	FE	PB	PART	ORG	SUL	NIT	CL	CO
MN	1.0000 (0) S=.001	-.0527 (454) S=.263	.7952 (430) S=.001	.0483 (456) S=.303	.8248 (455) S=.001	.0406 (451) S=.390	.1837 (453) S=.001	.1326 (453) S=.005	.2423 (453) S=.001	.0977 (366) S=.062
CU	-.0527 (454) S=.263	1.0000 (0) S=.001	-.1238 (428) S=.010	.2601 (454) S=.001	-.0067 (453) S=.887	.2619 (449) S=.001	.1147 (451) S=.015	.2302 (451) S=.001	.2268 (451) S=.001	.1949 (364) S=.001
FE	.7952 (430) S=.001	-.1238 (428) S=.010	1.0000 (0) S=.001	.1953 (430) S=.001	.7910 (430) S=.001	.1180 (425) S=.015	.2851 (428) S=.001	-.0572 (427) S=.238	.1840 (427) S=.001	.1771 (344) S=.001
PB	.0483 (456) S=.303	.2601 (454) S=.001	.1953 (430) S=.001	1.0000 (0) S=.001	.2284 (455) S=.001	.7883 (451) S=.001	.2753 (453) S=.001	.1409 (453) S=.003	.3734 (453) S=.001	.7097 (366) S=.001
PART	.8248 (455) S=.001	-.0067 (453) S=.887	.7910 (430) S=.001	.2284 (455) S=.001	1.0000 (0) S=.001	.2551 (450) S=.001	.3097 (453) S=.001	.1219 (452) S=.009	.3121 (452) S=.001	.2454 (366) S=.001
ORG	.0406 (451) S=.390	.2619 (449) S=.001	.1180 (425) S=.015	.7883 (451) S=.001	.2551 (450) S=.001	1.0000 (0) S=.001	.2373 (448) S=.001	.1584 (448) S=.001	.2618 (448) S=.001	.7167 (363) S=.001
SUL	.1837 (453) S=.001	.1147 (451) S=.015	.2851 (428) S=.001	.2753 (453) S=.001	.3097 (453) S=.001	.2373 (448) S=.001	1.0000 (0) S=.001	.1404 (450) S=.003	.2418 (451) S=.001	.2298 (364) S=.001
NIT	.1326 (453) S=.005	.2302 (451) S=.001	-.0572 (427) S=.238	.1409 (453) S=.003	.1219 (452) S=.009	.1584 (448) S=.001	.1404 (450) S=.003	1.0000 (0) S=.001	.0768 (450) S=.104	.0047 (364) S=.928
CL	.2423 (453) S=.001	.2268 (451) S=.001	.1840 (427) S=.001	.3734 (453) S=.001	.3121 (452) S=.001	.2618 (448) S=.001	.2418 (451) S=.001	.0768 (450) S=.104	1.0000 (0) S=.001	.2946 (363) S=.001
CO	.0977 (366) S=.062	.1949 (364) S=.001	.1771 (344) S=.001	.7097 (366) S=.001	.2454 (366) S=.001	.7167 (363) S=.001	.2298 (364) S=.001	.0047 (364) S=.928	.2946 (363) S=.001	1.0000 (0) S=.001

additional information listed. The number in parentheses in the center of each group of threes is the number of cases or patterns included in the correlation. The "$S =$" number is a statistical significance figure. The scientific meaning of these significance numbers may be difficult to decide. Truly "significant" S numbers depend on the types of data used. In some cases, a correlation of 0.80 may be considered weak, at best, if the variables used are expected to be highly related. In other cases where complex relationships exist between the variables, a correlation of 0.80 may indicate a remarkable relationship. The significance number cannot take this into consideration, and only represents a statistical calculation to test significance. User interaction is necessary to decide "significance" in a given data set. These numbers can, however, be used to compare the various relationships within a data set.

CORRELATION RESULTS

When studying the correlation values (the top number in the group of three) two types of correlations are interesting: (a) those with correlation values close to either $+1$ or -1 which means total relationship, and (b) those with values close to zero which show a total absence of linear relationship. Values midrange between these are difficult to interpret. The partial relationship seen may be a true sign of a small relationship, or may just be due to spurious data. Remember that this statistic gives no indication of strong, nonlinear relationships.

From Table II.3a, it is seen that MN is highly correlated with PART (0.82) and FE (0.79). The PB concentration is highly correlated with ORG (0.79) and CO (0.71). SUL, NIT, CL, and CU are not highly correlated with any other variables. This is shown by the midrange correlation values for these with all other variables.

Step II.3a should be continued to study whether these same relationships hold in all four subfiles (MC, MS, NS, and SC) individually, or if one or two are dominating the relationship. Also, other independent relationships may exist. This is done by running subfiles EACH.

STEP II.3b. BIVARIATE PLOTS

The next step, Step II.3b, can be used to plot interesting relationships found in the previous step. If the user finds a relationship that varies greatly between EACH or one that has *a priori* interest, the next step could be to study the relationship in more detail. A visual plot will give much more information than a single correlation coefficient for the relationship. Correlation coefficients indicate a match to a linear model only, and do not allow for other strong, but nonlinear relationships that could be visually seen by plotting. However, to plot all bivariate pairs will usually give an unmanageable amount of output. Therefore, Step II.3a is used first to identify those variables whose plots might be most interesting, keeping in mind that only linear correlations are assumed.

Each of the three major packages has plotting routines, and the choice depends on the characteristics desired for the plot. Often it is important to keep track of the location of each pattern point on a certain graph. Only ARTHUR has this ability. Using VAR-VAR in ARTHUR, one can request that (a) the case label be plotted, (b) the case name be plotted (user-defined), (c) the category number be plotted, or (d) any combination of the three above be separately plotted. For example, for our study we could plot case number, category number (1 for MC, 2 for MS, 3 for NS, and 4 for SC), or case name (some abbreviation of MC75JR12). This is useful for identifying outliers still present (most were eliminated in BMDP2D in Step II.1d) and other points of interest. Table II.3b Part A shows a typical FE versus PART plot from ARTHUR. Category numbers were chosen to be plotted.

At this point, one often is interested in trends between variables instead of individual points. In SPSS, the routine SCATTERGRAM can be used, and by running subfiles EACH And ALL, one can compare interesting trends seen from VARVAR in Part A of the table in more detail.

For instance, from Table II.3a, it was seen that FE and PART had a correlation of 0.79. This means that, in general, an increase in FE is accompanied by an increase in PART. This can be visually seen from the VARVAR plot. If this relationship is to be studied in more detail, a SCATTERGRAM plot can be made and is shown in Table II.3b Part B. As in the previous step, the file NONAME was used which contained subfiles MAR, MESA, NSCT, and SCOT (accomplished by running subfiles ALL). Values for the variable PART are plotted on the y axis and the FE values on the x axis. In this plot, an asterisk is used to denote the position of a single point. A numerical value plotted represents the number of cases positioned at that point. In VARVAR, this is not done. Only a single point is plotted at that position and the others are not included. Note that in Table II.3b Part A, the two far right columns tabulate the number of points plotted and not plotted at a given horizontal position. When plotting category numbers, as in this table, these columns should be checked. If a position contains two points, one for category 1 and one for category 2, the numeral "1" will be plotted since it was encountered in the data set first. The point belonging to category 2 will not be plotted but listed in the "NOT" column. The plot does not indicate whether the two points represent two different categories or not. It also does not designate, for points at a given horizonal position, which one of the positions represents a place occupied by more than one point. SCATTERGRAM does this by plotting the number of points there. It cannot, therefore, plot category numbers since it is using the plot symbols to represent number of points and not category membership for the point. This can only be done by requesting a separate plot for each category by designating a run of subfiles EACH.

The SCATTERGRAM plot was run after the extreme outlying point (PART = 589 and FE = 12.5) was removed. This point is seen in the extreme upper right corner of Table II.3b Part A. In SCATTERGRAM, limits on high and low values can be user-defined. This was done, and is shown at the bottom of the SCATTERGRAM plot as "EXCLUDED VALUES-1." Note that only 429 cases were plotted since 26 had either the value for PART or FE designated as "missing." The excluded value has caused the correlation coefficient to drop from 0.79 to 0.77. This indicates that both the FE and PART were

TABLE II.3b

Part A. VARVAR Plot of PART versus FE (from ARTHUR)

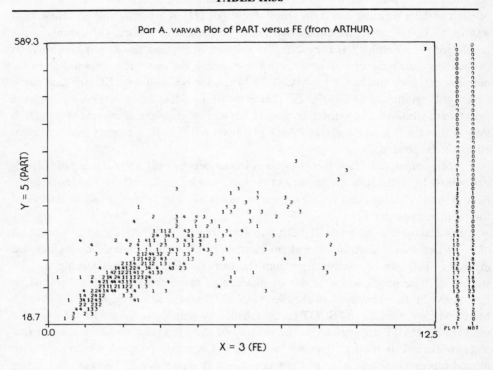

Part B. SCATTERGRAM Plot of PARTY versus FE

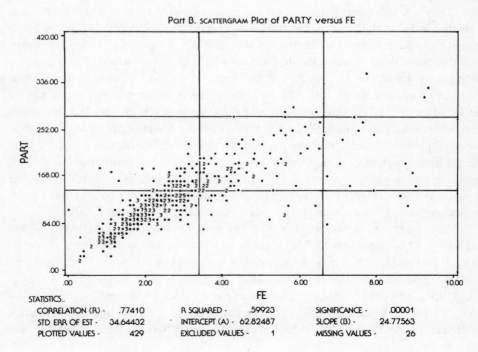

STATISTICS..

CORRELATION (R) - .77410	R SQUARED - .59923	SIGNIFICANCE - .00001
STD ERR OF EST - 34.64432	INTERCEPT (A) - 62.82487	SLOPE (B) - 24.77563
PLOTTED VALUES - 429	EXCLUDED VALUES - 1	MISSING VALUES - 26

TABLE II.3b *(Continued)*

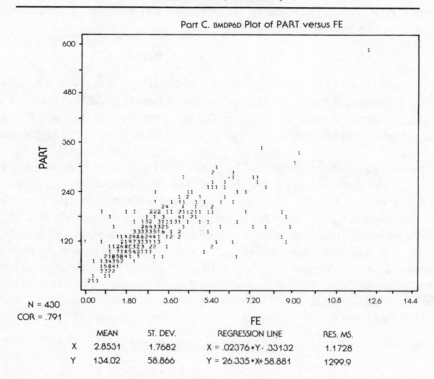

Part C. BMDP6D Plot of PART versus FE

	MEAN	ST. DEV.	REGRESSION LINE	RES. MS.
X	2.8531	1.7682	X = .02376•Y- .33132	1.1728
Y	134.02	58.866	Y = 26.335•X+ 58.881	1299.9

N = 430
COR = .791

equally extreme in this case. This can be seen in Part A since the point appears to lie on the line through the data. This fact is left to the scientist to explain.

The two-dimensional graph gives more feeling for the relationship than a single correlation coefficient. A general increase in FE with an increase in PART can be visually seen. From the plot, it may be suggested that the relationship holds better at lower values than at higher ones. Above a value of 6.50 for FE on the plot, the data become much more scattered. Below an FE value of 3.50, the relationship looks very strong. It would be instructive to run subfiles EACH to see if these trends are present for each category or not.

Bivariate statistics for the two-dimensional plot are listed. A least-squares line for the data is calculated to be

$$PART = 24.8 \ (FE) + 62.8$$

Note that the line is heavily influenced by the scatter above an FE value of 6.5. If only the lower points are considered visually, the intercept looks much closer to zero, with a steeper slope. The standard error of the estimate indicates that this equation can predict PART values from FE to about 34.6 units.

BMDP6D is also a useful plotting routine. In Table II.3b, Part C, the same plot of FE versus PART is shown, retaining the point that was excluded for SCATTERGRAM. Note that the line has changed to

$$PART = 26.3 \ (FE) + 58.9 \ \text{or} \ FE = .024 \ (PART) - .33$$

Both lines are listed. The correlation is again listed as 0.79 and the $N = 430$ indicates that 430 cases were again considered. Mean and standard deviation values for both variables are listed. BMDP6D will plot any size plots specified by the user. Either the number of points at each position (as is given in Part C) or category indications can be requested.

It should be noted that the size of the plots in SCATTERGRAM cannot be altered and the VARVAR plot sizes are variable only if a Calcomp plotter is used. In BMDP2D any size plot can be defined. Also, each axis can be increased independent of the other one to help in plotting one variable that has an extraordinary large number of values.

In summary, although correlations from Step II.3a show a numerical value that is related to correlations between variables, it is much more informative to see these relationships in a two-dimensional plot. To plot all possible pairs of variables initially would cause a large number of plots to be made, only a few of which may prove to be very interesting to the problem at hand. Therefore, the procedure in Step II.3a is utilized prior to II.3b to help indicate which plots are to be made for this step.

PARTIAL CORRELATIONS

The next procedure in studying intervariable relationships is to study partial correlations. A partial correlation gives a measure of linear association between two variables while statistically adjusting or controlling for the effects of one or more other variables. As assumption must be made that the effect of the control variables is linear.

First, the linear relationships between the independent, dependent, and control variable(s) are calculated. For instance, considering an independent variable A, a dependent variable B, and the control variable C, correlations of A with C, B with C, and A with B must be calculated. This A with B correlation allows C to be present as well.

Now, a new value of A and B must be calculated after removing the effects of C. This is done by calculating the difference between the actual value of A and the value of A predicted from the value C. For instance, using the relationship from SCATTERGRAM in Table II.3b Part B, let FE equal variable A and PART equal variable C. Considering a plotted point where $A = 5.0$ and $C = 160.0$, and assuming that we want to find the relationship between FE (A) and MN (B) controlling for PART (C), we must calculate a new value for A that controls for C. In other words, we must calculate the differences between the actual value of A and the value of A predicted from the value of C. The actual value of A is 5.0. The equation for the best line defining the relationships between A and C is given in the statistics at the bottom of Table II.3b Part B as $Y = mX + b$ or PART $= 24.8$ FE $+ 62.8$. In this case, the predicted value of FE $= (160 - 62.8)/24.8$ or 3.92. The actual value was 5.0; therefore the part of A *not* explained by C is $5.0 - 3.92 = 1.08$. This is the new A.

A similar procedure is done for *B* with *C*. Finally, a new correlation (called a partial correlation) of *A* with *B* is calculated using these new values of *A* and *B*. The effects of *C* have now been linearly statistically removed. This process is useful in uncovering spurious relationships (or correlations) by locating intervening variables. It also can find hidden correlations masked by the controlling variables.

Partial correlations can be used to control for the effects of more than one variable. Higher-order partials are computed from the previous ones. After the first-order partial is calculated, the linear effect of a second control variable can be statistically removed in a similar manner. This process continues until all specified control variables are included.

STEP II.3c. PARTIAL CORRELATIONS

For example, assume we want to further study the three-way relationship seen in Step II.3a between PART, FE, and MN using partial correlations. Both BMDP (BMDP6R) and SPSS (PARTIAL CORR) have programs for such a run. We feel SPSS is slightly easier to implement because multiple problems can be defined within a single computer run. In the PARTIAL CORR statement of SPSS, only a slash is necessary to define the second problem after the first one. In BMDP6R new regression paragraphs are needed for each, which is really not much more difficult. Also, with multiple control variables in SPSS, you can define the levels of inclusion whereas BMDP automatically prints the entire list. This means that each control variable is used individually, in every combination of two each, three each, etc., until all control variables (and combinations of these) have been considered. But it is felt that both programs give equally valid results. Both are quite similar. BMDP6R does give a more complete statistical study of the significance of the results.

Table II.3c gives a summary of results obtained for the above problem. Note that the values of simple correlations for the total data base are those from Table II.3a. All

TABLE II.3c
Partial Correlation Analysis

A. Simple correlations

	Total	MC	MS	NS	SC
MN–FE	.80	.60	.80	.94	.75
MN–PART	.82	.72	.93	.91	.86
FE–PART	.79	.74	.82	.90	.68

B. Partial correlations controlling for the variable in parenthesis

MN–PART (FE)	.67	.48	.80	.83	.72
FE–PART (MN)	.32	.56	.32	.15	.11

three values are quite highly correlated for the total data base. But note that when this is broken down into the four monitoring stations, the relationships in categories MC and SC are not as strong as in the other two. When the correlations for MN and PART are calculated, controlling for the influence of FE, all values drop somewhat with only MC dropping significantly. However, when the FE–PART correlations are calculated controlling for MN all values drop dramatically with MC dropping the least. This suggests that for MC, the real relationship probably exists mainly between FE and PART. For the other three categories, the real correlation is between MN and PART with the FE–MN and FE–PART relationship being defined by the MN.

Higher-order partial correlations can now be calculated. For example, the effects of both FE and ORG can be controlled in the PART versus MN relationship. It does not matter what order is chosen for control variable inclusion. On the printout from PARTIAL CORR, any or all of the order results may be specified. Zero-order correlations may be requested and significance testing of the results calculated.

The number of partial correlations possible is great. Therefore, the procedures in Steps II.3a and II.3b should be made before this is attempted to find which subgroups of related variables have potential interest and importance in the problem.

STEP II.3d. CLUSTERING OF VARIABLES

The next step in the process, Step II.3d, can either be used at this point or before Step II.3c depending on how well the relationships between variables are defined. In this step, BMDP1M is utilized to apply clustering methods to the data. Multivariate relationships between the variables are sought instead of simple bivariate ones as are found in Step II.3a and modified bivariate ones from Step II.3b. There are no equivalent techniques in SPSS. In ARTHUR, this can be done with HIER. In this program, however, you must first use CHANGE to redefine the variables (MN, PB, etc.) as cases and vice versa. Clustering is then applied to identify *similarities* (or differences) between variables and not between cases, as is usually done. The results from HIER are easier to read, but BMDP1M gives more output information regarding the results.

The user must first decide how to define a measure of association for clustering the variables. Some type of similarity definition is needed. Usually either correlation coefficients or Euclidean distances are used. To cluster variables, the common choice is correlations. When cases are clustered instead of variables, distance measurements are often the choice (see Step II.4b).

Initially, each variable is considered as an individual cluster. This means that each variable is distinctly defined by a point in n-dimensional space (where n is the number of cases in the data set). The correlation value which is used as the measure of association or similarity between each pair of clusters is calculated. This step involves calculating the correlation coefficients for all pairs of variables. The most similar clusters (or most similar variables in this instance) are noted. The user may define "most similar variables" in one of two ways. The first choice is to consider only those pairs of variables which have a large positive association (an increase in the first being reflected by an increase in the second) as being similar. The second choice is to consider those pairs of variables that have either a large positive or a large negative association, and therefore have correlation coefficients close to either positive or negative one. This is user-defined by choosing to either (1) include the signs of the coefficients, thus giving a scale from -1 to $+1$ as possible values or (2) ignore the signs and consider only the absolute value of the coefficient, thus giving a scale from 0 to $+1$.

After the most similar clusters are found, the next step is to link these together to form a new cluster. The user must now decide how to define the location of this new cluster in space, so that the similarity of this new cluster to all other clusters in the set can be calculated. BMDP1M allows three possible choices for this. The first one considers the correlation coefficient for two clusters to be the largest coefficient that is present between two points, one of which is present in each cluster. This is called single linkage clustering. If only a single point exists in each cluster, this becomes the correlation coefficient between this pair of points. If, however, one (or more) cluster(s) contain more than one point, the value chosen is the largest coefficient that exists between any pair of points, considering all pairs possible where one point lies in each of the two clusters. The second choice available is complete linkage where the correlation coefficient chosen is the smallest coefficient that exists between any pair of points, considering all possible pairs where one point lies in each of the two clusters. The third choice is average linkage where the value chosen is the average of the coefficients for all possible pairs of points between the two clusters.

After the linkage rule has been decided, new coefficients are calculated and that pair of clusters having the largest coefficient is then combined to form a new cluster. This pair could have consisted of either two single-point clusters, a single-point and a multiple-point cluster, or two multiple-point clusters. This process continues until all variables are included in a single cluster.

Results for the program are printed in a dendrogram. In BMDP1M, this dendrogram is referred to as a tree diagram. Pattern recognition definitions can become confused at this point, since the word "tree" is usually reserved for a different type of clustering technique. We will, therefore, refer to the figure as a dendrogram.

RESULTS OF THE CLUSTERING

Table II.3d illustrates the results of the variable clustering procedure from BMDP1M, using the total data set. The default of single linkage clustering was used. It would be useful to compare this to ones obtained from complete and average linkage clustering. On the left is the dendrogram. The ten variables are linked vertically with the variable number in parentheses. This number indicates their order of appearance in the data base and not in the dendrogram. The dendrogram is shown with the numbers on it representing scaled versions of the correlation coefficient values. A key to this scale is given in the upper right of the table. The program gives the user a choice of scales to be used. A value on the dendrogram between 0 and 5 represents a correlation coefficient between -1.0 and -0.9. A value between 5 and 10 represents a coefficient between

TABLE II.3d
BMDP1M Clustering Analysis for the Variables

TREE PRINTED OVER CORRELATION MATRIX (SCALED 0-100).   THE VALUES IN THIS TREE HAVE BEEN SCALED 0 TO 100
CLUSTERING BY MINIMUM DISTANCE VALUE.   ACCORDING TO THE FOLLOWING TABLE

VALUE ABOVE	CORRELATION		VALUE ABOVE	CORRELATION
0	-1.000		50	.000
5	-.900		55	.100
10	-.800		60	.200
15	-.700		65	.300
20	-.600		70	.400
25	-.500		75	.500
30	-.400		80	.600
35	-.300		85	.700
40	-.200		90	.800
45	-.100		95	.900

NAME	VARIABLE NO.	OTHER BOUNDARY OF CLUSTER	NUMBER OF ITEMS IN CLUSTER	DISTANCE OR SIMILARITY WHEN CLUSTER FORMED
MN	2	9	10	64.47
PART	6	2	2	91.92
FE	4	2	3	87.98
PB	5	10	4	76.12
ORG	7	5	2	89.24
CO	11	5	3	85.75
CL	10	2	7	70.34
SULF	8	2	8	66.88
CU	3	2	9	65.96
NIT	9	2	10	64.47

```
VARIABLE
NAME    NO.
MN    (  2) 91/87/63 61 56 65/61/50/55/
PART  (  6)/87/69 70 63 69/66/52/56/
FE    (  4)/65 60 58 61/64/45/46/
PB    (  5) 89/85/76/65/61/57/
ORG   (  7)/85/69/61/62/57/
CO    ( 11)/67/60/59/48/
CL    ( 10)/62/65/59/
SULF  (  8)/55/57/
CU    (  3)/64/
NIT   (  9)/
```

AN EXPLANATION OF THE VARIABLE CLUSTERING PROCESS SHOWN IN THE TREE PRINTED ABOVE

THE PROCESS BEGINS WITH THE CLUSTER CONSISTING OF VARIABLE PB (5), THE 4TH VARIABLE LISTED IN THE TREE.
THIS CLUSTER JOINS WITH THE CLUSTER BELOW IT CONSISTING OF VARIABLE ORG (7).
THE NEW CLUSTER IS INDICATED ON THE TREE BY THE INTERSECTION OF THE DASHES BEGINNING ABOVE VARIABLE PB (5)
WITH THE SLASHES STARTING NEXT TO VARIABLE ORG (7).

THIS CLUSTER JOINS WITH THE CLUSTER BELOW IT CONSISTING OF THE VARIABLE CO (11).
THE NEW CLUSTER IS INDICATED ON THE TREE BY THE INTERSECTION OF THE DASHES BEGINNING ABOVE VARIABLE PB (5)
WITH THE SLASHES STARTING NEXT TO VARIABLE CO (11).
THE PROCESS CONTINUES UNTIL EACH VARIABLE IS JOINED TO AT LEAST ONE OTHER VARIABLE.

−0.9 and −0.8 and so on. This key, then, is chosen to define a high correlation as having a large positive association, when an increase in one variable is accompanied by an increase in the second one. A highly negative correlation, in this scale, is keyed to have a low association. The user may choose to define a high key value to indicate a strong association between variables whether it is positive or negative.

A summary of the clustering procedure is listed below the key. The second line indicates that PART, variable number 6, was added to the cluster bounded by variable number 2 (MN) to become a two-item cluster. This occurred at a similarity value of 91.92. This is shown on the first two lines of the dendrogram. Note also that ORG (variable 7) was combined with variable 5 (PB) to make a two-item cluster at a similarity of 89.24. This can also be seen on the dendrogram. In the next step CO (variable 11) was added. The table indicates that CO was combined with the cluster bounded by PB (variable 5) to make a three-item cluster at a similarity value of 85.75. The bottom part of the table explains these first few steps. The process continues until all variables are considered to be a single cluster.

This procedure, Step II.3d, is used to find the relationship between variables. Depending on what degree of association is acceptable, the variables may be clustered into any number of groups. Above a keyed value of 92, each variable is an individual cluster. Below a keyed value of 64, all variables are contained in a single cluster. Above a keyed value of 75, MN, PART, and FE form a cluster as do ORG, PB, and CO. Clusters of this type can be very useful for Step II.3c. Partial correlations between these variables can be studied. Results are also useful for input into Step II.3e where correlations are then studied between groups of variables.

The dendrogram from HIER in ARTHUR (shown later in Step II.4b) is easier to read. An extra step, however, is necessary before considering the variables as the cases or transposing the matrix. The output does not give a summary key. BMDP1M was therefore chosen for ease of implementation.

It would be instructive to compare variable groupings obtained for each county individually. A comparison of these can give insight into the cause of certain variable values.

CANONICAL CORRELATIONS

In the next step, Step II.3e, canonical correlations are studied, which are a type of combination of multiple regression and factor analysis. The relationship between two groups of variables is sought. Each group must contain variables that are interrelated in some way, and can be given a theoretical meaning as a set. An example would be to study the relationship between a group of weather variables and a group of pollution variables. This is done in the following manner. A linear combination equation within each set of variables is created which maximizes the linear correlation between that set of variables and a similar linear equation found for the second set of variables. The maximum amount of relationship between the two sets is sought.

The number by which a given variable is multiplied in the linear combination created is called that variable's canonical coefficient or loading. The first pair of canonical variate sets is created to have the highest intercorrelation possible. The second set of variates is then chosen that accounts for the largest amount of *remaining* relationship (that which has not been accounted for in the first set). The second set of variates is uncorrelated with the first set. The process continues in this manner. Significance testing for the relationship can be done.

Canonical correlations are available in either CANCORR (from SPSS) or BMDP6M. Both give similar results. No routine for this is available in ARTHUR, although an ARTHUR-compatible program, PLS-1, is available from the ARTHUR source, Infometrics, Inc., that will do canonical correlations (and more). This program will be discussed later. In the example used here, BMDP6M results are shown.

The most difficult step in utilizing canonical correlations, in most cases, is to define the two groups of variables. To successfully apply this technique, it must be assumed that the variables included within each set are related to each other by some underlying principle. Also, a relationship must exist between the two sets. These sets are not often easy to determine, but performing Steps II.3a through II.3d may help in separating the variables into the groups to be used.

In some problems, variable groupings may be inherent in the original variable choices. For our example, we want to determine what may be affecting the pollution measurements studied to this point. One choice for this would be to add to the data base some weather variables. A relationship between the variables which represent weather conditions and those measuring pollutants could then be made.

STEP II.3e. APPLICATION OF CANONICAL CORRELATIONS

Weather variables were added to the data set, and results of the canonical correlation program from BMDP6M are given in Table II.3e. Part A of the table indicates the weather variables that were added. The data for these variables were obtained from the Sky Harbor International Airport in Phoenix, which is located relatively close to the MC sampling site. In this case, only those data for pollution from the MC site were used, and the relationship of these to the weather data was sought.

Many printout options are available in BMDP6M. Univariate summary statistics for each of the variables in the data set can be listed. These include means, standard deviations, the smallest and largest values, and the skewness and kurtosis.

The two subsets of variables are considered. In Part B of Table II.3e, the two subsets shown include variables VISIB to RELHUM and MN to CO. It does not matter in which order these variables are considered. A value (the R-SQUARE value) is listed which indicates the correlation of each variable in the set with all other variables in that set via multiple linear regression. For instance, in the weather data set (first set), DEWPT has a square multiple correlation of 0.96 with the other variables in that set.

TABLE II.3e
Canonical Correlations of Pollution and Weather Data

Part A   Weather Variables

VISIB- visibility	DEWPT- dew point	MDWDSP- mode wind speed
TEMP- temperature	SOLRAD- solar radiation	PKGUST- peak gust
SUNCOV- sun cover	WDSP- windspeed	RELHUM- relative humidity
	PPT- precipitation	

Part B

SQUARED MULTIPLE CORRELATIONS OF EACH VARIABLE IN THE SET WITH ALL OTHER VARIABLES IN THAT SET

VARIABLE

NUMBER	NAME	R-SQUARED	NUMBER	NAME	R-SQUARED
16	VISIB	.06376	2	MN	.49119
17	TEMP	.96663	3	CU	.34401
18	SUNCOV	.42439	5	FE	.68514
19	DEWPT	.96325	6	PB	.83560
20	SOLRAD	.73928	7	PART	.75390
21	WDSP	.68256	8	ORG	.86970
22	PPT	.16207	9	SUL	.25073
23	MDWDSP	.60456	10	NIT	.34224
24	PKGUST	.12272	11	CL	.35150
25	RELHUM	.93977	12	CO	.79140

Part C   Part D

EIGENVALUE	CANONICAL CORRELATION	NUMBER OF EIGENVALUES	BARTLETT'S TEST FOR REMAINING EIGENVALUES		
			CHI-SQUARE	D.F.	SIGNIFICANCE
.82892	.91045	0	311.11	100	.00000
.55726	.74650	1	174.28	81	.00000
.39173	.62588	2	111.13	64	.00024
.36088	.60073	3	72.61	49	.01587
.25029	.50029	4	37.91	36	.38218
.12698	.35635	5	15.59	25	.92658
.05228	.22865	6	5.06	16	.99544
.00925	.09619	7	.90	9	.99963
.00224	.04731	8	.18	4	.99617
.00009	.00931	9	.01	1	.93468
		10	.00	0	1.00000

BARTLETT'S TEST ABOVE INDICATES THE NUMBER OF CANONICAL VARIABLES
NECESSARY TO EXPRESS THE DEPENDENCY BETWEEN THE TWO SETS OF VARIABLES.

Part E

CANONICAL VARIABLE LOADINGS

		F1	F2	F3			S1	S2	S3
MN	2	.265	.330	.335	VISIB	16	-.105	-.584	-.497
CU	3	.242	.093	.061	TEMP	17	-.667	.134	-.139
FE	5	.244	.473	.136	SUNCOV	18	.219	-.432	.446
PB	6	.916	.219	-.025	DEWPT	19	-.473	-.137	-.346
PART	7	.316	.546	.460	SOLRAD	20	-.668	-.201	.286
ORG	8	.902	.256	.046	WDSP	21	-.797	.058	.238
SUL	9	.159	.578	-.232	PPT	22	-.201	.052	-.347
NIT	10	.091	-.184	.704	MDWDSP	23	-.558	.019	.245
CL	11	.332	.357	-.099	PKGUST	24	-.142	.075	.155
CO	12	.808	.518	.003	RELHUM	25	.208	-.396	-.182

The usefulness of these values is in optimizing the number of variables in each set. Since DEWPT is readily predicted from the other nine variables in this group (as shown by its high correlation value), the information contained in it is somewhat redundant, and could be predicted from the other variables. Care must be exercised in variable eliminations, however, and the R^2 value of the new set created by the elimination of the variable should be compared to the original value. You do not want to eliminate a variable that has some importance. Likewise, check TEMP and RELHUM, and possibly ORG and PB in the pollutants data. The R-SQUARE value will be discussed in detail in Step II.3f.

The next procedure determines the sets of canonical variates that represent relationships between the pollution and weather data. Results are given in Part C of the table. The first set created by the program has a correlation of 0.910 (or an eigenvalue of 0.828 which is the square of the correlation). A second set of canonical variates is then determined which accounts for the largest amount of *remaining* variance (that not described in the first set). This set has a correlation value of 0.747 (or an eigenvalue of 0.557). This process continues until all the variance is described. However, usually only the first few sets of canonical variates are significant. Bartlett's test can be used as a test of significance at this point in BMDP6M (see Part D of the table). In CANCORR in SPSS, other tests of significance, such as Wilk's lambda, are available.

In Table II.3e, Part D, the value listed under "SIGNIFICANCE" determines how well the problem can be described by the eigenvectors above it. A value of zero means that not all of the variance has been described and a value of 1.00 means that all of the variance has been described. Therefore, in this case, after the first step ("NUMBER OF EIGENVALUES" equals "1") the significance value listed equals zero. This means that more than one set of variates is necessary to describe the data. After the second step, the significance value goes to 0.00024, which still indicates that more sets of variates are necessary. After five steps, the significance value becomes 0.927. This indicates that although ten sets of variates are determined, only the first five are truly significant. The last five sets of canonical variates could be eliminated without much loss of information. The sixth could have been included to bring the value up to 0.995.

The canonical variable loadings are then listed in Part E of Table II.3e. Only the first three sets are shown. These can be thought of as representing the importance of each variable in this set. Therefore, the first set of canonical variates (listed under CNVRF1 for the pollution data and CNVRS1 for the weather data) has a canonical correlation of 0.91. This set loads high in PB, ORG, and CO in the pollution data and TEMP, SOLRAD, and WDSP in the weather data (also notice the negative values). A preliminary evaluation of this result would be that increasing the temperature, solar radiation, or windspeed would decrease the PB, ORG, and CO. It is now up to the experimentor or investigator to give this scientific meaning. This is a step the computer cannot perform. The results here are not unexpected since pollution is known to be less prevalent in Phoenix in the summer (high temperature and solar radiation) and it is expected that winds would also blow these pollutants out of the valley. This type of interpretation can be continued for the other sets of variates.

Finally, a plot can be made of these two linear combinations of variables to study the relationship in more detail. This plot will give a better visual indication of the relationship than a single correlation value can, similar to Steps II.3a and II.3b.

Other optional outputs for BMDP6M are the correlation matrix, covariance matrix, plots of the original variables, and listing of the original data. It is critical in this step of the procedure to carefully choose meaningful subsets of the data if meaningful results are to be obtained. Results from CANCORR in SPSS are similar.

REGRESSION ANALYSIS

The final section of Step II.3 is Step II.3f, a method for doing regression analyses of various types. The goal of regression analyses is to study the relationship between a given variable (called the *dependent variable*) and a set of *independent variables*. Often it is desirable to find an equation of the independent variables that will best predict the value of the corresponding dependent variable. Many types of regression approaches are available. BMDP has a variety of programs. BMDP6D does a simple linear $y = mx + b$ fit for a single dependent and single independent variable. BMDP1R does a simple multiple regression. In this case the entire set of independent variables is automatically included in the analysis. A linear combination of these is then found that best predicts the value of the dependent variable. Given independent variables A, B, C, and D and dependent variable X, an equation of the form

$$X = aA + bB + cC + dD + e$$

is found that best predicts values of X.

BMDP2R performs a stepwise regression analysis. Variables are considered individually and are added or removed from the equation in a stepwise manner. Four criteria are available to determine which variables are to be included (or excluded). Results of the final equation obtained with this method are similar to those with BMDP1R. However, the intermediate steps are available here. Evaluations of results are made after each step. Both methods assume only linear relationships.

BMDP3R performs a nonlinear regression. A choice of five different commonly used nonlinear functions is available to the user. Other functions can be fitted to the data by specifying both the function and its derivatives by FORTRAN statements. BMDP5R is also a nonlinear regression analysis which assumes a polynomial equation of the form $y = B_0 + B_1x + B_2x^2 + \cdots + B_px^p$.

Two new regression programs have been introduced to BMDP. BMDP9R studies "best" subsets of the predictor variables. Various criteria to define "best" are present. The number of best subsets (up to ten) can be chosen. This shows the prediction ability using various combinations of variables. Regression results never have a single correct answer and this routine studies some of the other ones available. BMDPAR is also a new addition. This routine studies derivative-free nonlinear regressions. The user supplies the function to which the data will be fitted. This is similar to BMDP3R but the derivative of the function need not be specified here. A BMDP technical report* is available which gives an excellent detailed explanation of interpretations from BMDP2R and BMDP9R.

In SPSS, a single program, REGRESSION, is used and options dictate whether a fixed number of variables, a stepwise technique, or a combination of these is used. No nonlinear or polynomial regressions are available. But since in most scientific applications, the true function of the relationship is usually not known, this disadvantage is minimal. However, there is no SPSS equivalent program to the very useful BMDP9R. Therefore, most use will be made of BMDP2R, BMDP9R, and possible BMDP1R. The BMDP programs do residual plots more conveniently and an option does exist to include a zero intercept as well. There are no nonlinear or polynomial regressions in SPSS.

It must be realized that when applying regression techniques, the solutions obtained are only good for that data set. Adding or deleting samples from the same population may result in quite different "optimal" solutions. The alternative solutions obtained from BMDP9R can be used to help evaluate the uniqueness of a solution.

The most difficult step in a regression analysis is to determine which variables should be included. The computer is a minimal aid in this decision, but results from Steps II.3a to II.3e will help. Also, with stepwise analyses, the best predictor variables can be chosen from a variable set. However, the results are extremely dependent upon the variable set chosen. Often, Step II.1d may suggest that transformations (such as logarithmic changes) be made first which will also greatly affect the regression results.

STEP II.3f. REGRESSION ANALYSIS

Table II.3f presents results from a regression analysis from the pollution data. It is divided into two parts. The first (Part A) gives a partial printout from BMDP1R. The problem attempts to predict SOLRAD from pollution data. An *a priori* knowledge of which pollutants were most important was not available so all were included. The same problem was then approached with REGRESSION in SPSS, but variables were stepped in one by one. The stepping criterion used was an increase in B value (explained later). Part B of Table II.3f shows results from the first two steps of this procedure. Similar printouts result after each step. The last part of the table shows a summary table from the stepwise regression after either (a) all variables were included or (b) those not included were below some certain defined tolerance level.

Looking first at Part A of Table II.3f (obtained from BMDP1R), only final results are given. Dependent variable 20 (SOLRAD) was predicted from the pollution data. The multiple R value listed is a correlation value between the dependent variable and the linear combination of independent variables that best fits the data. It can be interpreted in a manner similar to the simple correlation coefficient in Step II.3a. The multiple R-squared value (calculated by squaring 0.694) indicates the amount of variation

---

*BMDP Technical Report No. 48 *Annotated Computer Output for Regression Analysis* by Mary Ann Hill. See Appendix V for ordering information.

in the dependent variable that is explained by the equation. It should be noted that this value only indicates the amount and statistically tends to be overoptimistic. This is further explained in the BMDP technical report mentioned previously. The value in Part A (0.694 squared equals 0.481) indicates that 48% of the variation in the solar radiation is described by this equation. The standard error of the estimates is the absolute (in correct units) error to expect from the prediction. Therefore, values of solar radiation can be predicted ± 123 radiation units. The analysis of variance shows that the regression equation defined contains only about one-half of the sum of squares present. 1,101,050 was explained. The other 1,186,898 unexplained portion remains in the residual. An F-ratio can be used to study the significance of this. Details are given in Step II.2.

Now, to find the equation that was determined, the coefficient column is used as front numbers on the equation. Therefore, the following is obtained:

$$\text{SOLRAD} = 506 - 262(\text{MN}) - 24.9(\text{CU}) - 14.7(\text{FE}) \ldots$$

Therefore, if chemical analyses were obtained on a given day at station MC, measurements could be substituted into the above equation to predict values for the expected solar radiation for that day. The prediction would be correct (on the average) to within 123 solar radiation units.

TABLE II.3f
Regression Analysis

Part A Simple Regression from BMDP1R

```
REGRESSION TITLE. . . . . . . . . . . . . . .COUNTY DATA
DEPENDENT VARIABLE. . . . . . . . . . . . .     20 SOLRAD
TOLERANCE . . . . . . . . . . . . . . . . .  .0100
ALL DATA CONSIDERED AS A SINGLE GROUP

MULTIPLE R              .6937        STD. ERROR OF EST.       123.3558
MULTIPLE R-SQUARE       .4812
```

ANALYSIS OF VARIANCE

	SUM OF SQUARES	DF	MEAN SQUARE	F RATIO	P(TAIL)
REGRESSION	1101050.734	10	110105.073	7.236	.00000
RESIDUAL	1186898.344	78	15216.645		

VARIABLE		COEFFICIENT	STD. ERROR	STD. REG COEFF	T	P(2 TAIL)
(CONSTANT		505.6070)				
MN	2	-262.334	424.359	-.071	-.618	.538
CU	3	-24.927	11.831	-.212	-2.107	.038
FE	5	-14.660	15.584	-.137	-.941	.350
PB	6	-8.939	17.593	-.102	-.508	.613
PART	7	1.410	.522	.444	2.702	.008
ORG	8	-9.497	5.632	-.381	-1.686	.096
SUL	9	-1.944	5.223	-.035	-.372	.711
NIT	10	4.986	9.672	.052	.516	.608
CL	11	-.719	5.926	-.012	-.121	.904
CO	12	-.014	.010	-.251	-1.407	.163

(Continued overleaf)

TABLE II.3f (*Continued*)
Regression Analysis

Part B   Forward Selection Regression from REGRESSION (SPSS)

Step #1

* * * * * * * * * * * * * * * * * *   M U L T I P L E   R E G R E S S I O N   * * * * * * * * * * * * * * * * * *

DEPENDENT VARIABLE..   SOLRAD

VARIABLE(S) ENTERED ON STEP NUMBER 1..   ORG

MULTIPLE R	.59762	ANALYSIS OF VARIANCE	DF
R SQUARE	.35716		
ADJUSTED R SQUARE	.34977		
STANDARD ERROR	130.02196		

	DF	S UM OF SQUARES	MEAN SQUARE	F
REGRESSION	1.	817152.86892	817152.86892	48.33591
RESIDUAL	87.	1470796.74906	16905.70976	

- - - - - VARIABLES IN THE EQUATION - - - - -

VARIABLE	B	BETA	STD ERROR B	F
ORG	-14.89812	-.59762	2.14287	48.336
(CONSTANT)	572.36267			

- - - - - VARIABLES NOT IN THE EQUATION - - - - -

VARIABLE	BETA IN	PARTIAL	TOLERANCE	F
MN	.14626	.17258	.89506	2.640
CU	-.14140	-.16589	.88486	2.434
FE	.12552	.14896	.90543	1.952
PB	-.13654	-.08000	.22070	.554
PART	.30392	.32172	.72034	9.929
SUL	.00062	.00072	.88068	.000
NIT	.10600	.13082	.97919	1.497
CL	.04657	.05197	.80056	.233
CO	-.24951	-.16256	.27286	2.334

Step #2

DEPENDENT VARIABLE. .   SOLRAD

VARIABLE(S) ENTERED ON STEP NUMBER   2. .   PART

MULTIPLE R   .65091
R SQUARE   .42369
ADJUSTED R SQUARE   .41029
STANDARD ERROR   123.82322

ANALYSIS OF VARIANCE

	DF	SUM OF SQUARES	MEAN SQUARE	F
REGRESSION	2.	969381.38406	484690.69203	31.61262
RESIDUAL	86.	1318568.23392	15332.18877	

------ VARIABLES IN THE EQUATION ------

VARIABLE	B	BETA	STD ERROR	F
ORG	-18.90465	-.75834	2.40443	61.818
PART	.96485	.30392	.30621	9.929
(CONSTANT)	471.37589			

------ VARIABLES NOT IN THE EQUATION ------

VARIABLE	BETA IN	PARTIAL	TOLERANCE	F
MN	-.04737	-.04540	.52937	.176
CU	-.11120	-.13686	.87304	1.623
FE	-.12692	-.11344	.46039	1.108
PB	-.25780	-.15626	.21172	2.127
SULF	-.07268	-.08699	.82558	.648
NIT	.06903	.08899	.95785	.679
CL	-.04815	-.05391	.72249	.248
CO	-.22871	-.15723	.27236	2.155

Summary Table

VARIABLE	MULTIPLE R	R SQUARE	RSQ CHANGE	SIMPLE R	B	BETA
ORG	.59762	.35716	.35716	-.59762	-9.49660	-.38095
PART	.65091	.42369	.06653	-.09711	1.41009	.44416
CO	.66177	.43794	.01425	-.57769	-.01413	-.25130
CU	.67697	.45829	.02036	-.32791	-24.92743	-.21216
FE	.68865	.47424	.01585	-.07014	-14.66007	-.13672
PB	.69030	.47651	.00227	-.55770	-8.93928	-.10220
MN	.69181	.47861	.00209	-.06268	-262.33412	-.07068
NIT	.69292	.48014	.00153	.01757	4.98618	.05184
SUL	.69364	.48114	.00100	-.20589	-1.94424	-.03507
CL	.69371	.48124	.00010	-.22961	-.71872	-.01228
(CONSTANT)					505.60698	

STEPWISE REGRESSION

Comparing this to the stepwise procedure listed in Part B of Table II.3f, it can be seen that the same results are obtained, and similar statistics shown. The stepwise procedure, however, gives results after each step (or in other words, after each variable is added). This part was taken from REGRESSION in SPSS. SOLRAD was again used as the dependent variable. The pollution variables were added stepwise to the problem to form the regression line.

In SPSS, "B" is used as the symbol for the regression coefficient. Looking at Step 1, the single best variable to define solar radiation is ORG. At that point in the analysis, the multiple R is 0.597, 35.7% of the variance is described, and the equation used is

$$SOLRAD = 572 - 14.9\,(ORG)$$

A prediction can be made to within 130 solar radiation units. The variables not yet in the equation are listed on the right. A value for "PARTIAL" indicates the partial correlation or the degree to which a variable accounts for the remaining variation unaccounted for by the variables already in the equation. Notice that the highest value present after the first step is 0.30 for variable PART. Therefore, the second variable that would be included would be PART. Table II.3f, Part B, shows this second step as well. Note that it was PART that was chosen. The R square value now indicates that 42% of the variance is included and the predicting equation after this second step is

$$SOLRAD = 471 - 18.9\,(ORG) + 0.96\,(PART)$$

Again, the statistics for the "VARIABLES NOT IN THE EQUATION" are listed on the right. The next variable to be entered is that one having the highest "PARTIAL" value. Therefore, Step 3 should include the variable CO with a partial correlation after the second step of 0.157.

In the stepwise procedure, the process can continue (as in this case with REGRESSION) until all variables are included in the analysis. Results are listed after each step, however, so that the equation containing any given number of variables is obtainable. The second choice is to cease adding variables when the next variable would not significantly add to the prediction ability. This point can be defined in a variety of ways. Usually, it is determined by the F value listed for each variable not yet in the equation. In REGRESSION, the analysis quits if no variable has an F value greater than 0.01. This can be user-changed. In BMDP2R, the point is determined to be where the R squared value exceeds $(1.0 - TOLERANCE)$ where the "TOLERANCE" value defaults to 0.01 and can be user-changed.

A backwards stepping can also be chosen with REGRESSION or BMDP2R. In this case, all variables are originally considered to be in the equation and are removed according to some defined criteria. This continues until all are removed or some defined removal statistic is exceeded.

It is also possible to distinguish between variables that must be included in the equation and those that can be if prediction ability is increased. In this procedure, certain variables are forced into the equation. Both REGRESSION and the BMDP programs can do this.

The summary table for the stepwise procedure from REGRESSION is given at the bottom of Table II.3f. Variables are listed in the order of their inclusion. The multiple R (and R squared) tells how these values increase after each step is completed. Remember that these values tend to be overoptimistic. BMDP corrects for this somewhat by calculating either an adjusted R squared or a statistic called Mallows' C_p. The "RSQ CHANGE" indicates how R squared changes after each step. The simple R is a simple correlation coefficient between solar radiation and that variable. These are the same values as obtained from CORREL, Step II.3a. The "B" value is identical to those obtained from the simple multivariate linear regression problem (shown in Part A of the table). These are used to obtain the regression equation.

Nothing has been said to this point about BETA, which is a standardized regression coefficient. Appendix IX discusses standardized scores. If these are used, the value cannot be used to obtain regression equations as shown above since the effects of the different units on each of the variables have been taken into consideration. What a BETA value does indicate is the number of standard deviation units change in SOL-RAD expected when that given independent variable changes by one standard deviation unit. Therefore, for the first step of the regression analysis, SOLRAD changes -0.6 standard deviation units for each standard deivation unit change in ORG. In Part A of the table, the column titled "STD REG COEFF" lists the same BETA values.

SUMMARY

In summary, from the above results it is seen that a regression analysis is appropriate to define a dependent variable with a linear combination of independent variables. It is suggested that BMDP2R (or the stepwise REGRESSION) be chosen since final results are identical to simple regression but results after each step can be seen. Also, when the R-square change does not change significantly, it will be assumed that all necessary vari-bles have been included. If the relationship is not linear but is known, BMDPAR is used and similar results are obtained. Usually, however, the relationship is not known and the best that can be done is to assume linearity or to try some of the functions available in BMDP3R.

PLS-2

An interesting, new ARTHUR-compatible program combining many of the methods in this section is available from Infometrics, Inc. The program, PLS-2, is a partial least-squares path modeling program that is used to study sets of variables. The user can segregate the variables into appropriate blocks, and can determine the influence of each block in the overall model. If a two-block model is used with one of them containing a single variable, a multiple regression analysis results. If a two-block model is used with each block containing several variables, canonical correlation analyses result. PLS-2 allows this to be expanded to a multiple number of blocks, where interrelationships between these are studied to create a complex path model for the system. The program is unique in the extensive powerful ability allowed for studying intervariable relationships.

NONPARAMETRIC STATISTICS

Both SPSS and BMDP have the ability to do nonparametric statistics for correlations, t-tests, and analyses of variances. This is applicable to grouped variable data or data with a large number of groups where order has meaning (i.e., if MN values were divided into groups of 0–5%, 5–10%, 10–15%, etc.). Spearman or Kendall rank–order correlation coefficients can be calculated in the SPSS program NONPAR CORR. In BMDP3S, these are possible along with Wilcoxon sign-ranked, Friedman two-way, and Mann Whitney U-test methods. None are available in ARTHUR. A discussion and example of nonparametric statistical analyses is given in Appendix VII.

After successful completion of Step II.3, one should have a very thorough feeling for the variables chosen and how they relate to each other. The next step involves unsupervised learning techniques. This is most valuable for defining "natural" groupings in the data base considering all the variables simultaneously (or some subgroups of variables as defined by Step II.3). Also, even if groupings are known *a priori,* unsupervised learning will be completely unbiased in checking the validity of such groups and group differences.

II.4. UNSUPERVISED LEARNING TECHNIQUES

INTRODUCTION

Often at this point in a statistical analysis, data grouping may be studied in greater detail. This was discussed to a certain degree in Step II.2. However, in that case, variables were studied singularly, or at most a few at a time. Often it is desirable to be able to study the entire data base as a whole with no prejudices as to which variables are important or how group definitions are to be made. In the air pollution example, this is not an important manipulation. The four possible categories are already defined for us, and comparisons between these are desired. However, with many types of data, this step can be very useful for either studying the "natural" groups that are present in a data set or to check the previously defined categories in an unbiased manner. An example might be the study of lunar rock types. Before collection, no one was sure of just how many varieties of lunar rocks existed. The approach was to collect samples. Chemical and mineralogical analyses were then made on these. Some *a priori* prejudices as to what might exist were present and tentative groupings could be made, but a study of the simultaneous combination of all data available on the samples was necessary to "group" certain samples together. By using unsupervised learning techniques, these natural groupings could be studied without prejudice.

In Step II.4 no certain order for the steps is needed. Any of the three may be performed first. Each gives unsupervised learning information quite complementary to the other techniques. Therefore, it is only through a study of the results from a combination of these technqiues that meaningful classification results can be obtained.

CLUSTER ANALYSIS

Studies of regularities among objects is generally referred to as cluster analysis. A large number of studies have been done in this field of statistics recently, especially for applications in the sciences.

To understand this type of multivariate statistical analysis, the data structure is considered in the following manner. Assume the data consist of two variables measured for ten objects. The data can be thought of as a plot of the two values for each object in two-dimensional space. This can easily be done manually using the x axis for the first variable and the y axis for the second. Ten points would then be plotted. One could visually study the plot, looking for natural groupings present. To do this, some type of mental "similarity" definition would have to be used. A Euclidean distance is often used for this definition, and those points closest together (having the smallest distances between them) are designated as a "group." The number of groups chosen is arbitrary. As the requirements for group membership are relaxed, more points can be included in a group. Two groups may then be combined to gain simplicity while losing information. This type of analysis, therefore, gives an indication of "similarity" as related to distance or nearness.

When three variables are considered simultaneously, similar manipulations can be made. Although three-dimensional plots are difficult, the variables may be visually seen by plotting either contours of the three dimensions or plotting two variables at a time. Although harder to visualize, distances are still as valid in three-dimensional space as in two, and groups can be defined by "similarity" or distances as before. However, the job becomes increasingly difficult and the aid of a computer would be very useful.

Now consider "n" variables measured on "m" different objects. This could be thought of as "m" points plotted in n-dimensional space. Although this space cannot be manually plotted and visualized, distances are still meaningful as is closeness or similarity. Unsupervised learning can be used to study natural groupings in this n-dimensional space. Various definitions for "similarity" and for "group" are possible.

DATA SET CONSIDERED

In this step, a subset of the original data base given in Appendix III is used. Only the summer data (June, July, and August) were considered. The analyses were performed considering only the Maricopa County (MC) and North Scottsdale (NS) sites.

The goal was to study any natural groupings that might be present in the data. It would be interesting to determine if all of the June pollution data for both sampling sites were similar. The two groups may or may not retain the division along station lines when "natural" groups are sought. Some complex relationship between cases might be present. A possibility might also be that six distinct groups representing each station during each month exists, or that some of these may really be similar enough to combine, forming a smaller number of groups.

STEP II.4a. MINIMAL SPANNING TREE

In Step II.4a, a subroutine TREE in ARTHUR is used to study a minimal spanning tree. In this case, the data are considered as m points in n-dimensional space which represents m number of samples. A definition of "closeness" or distance is necessary. Usually, for TREE, an n-dimensional Euclidean distance is used. The distance between a given point and every other point in the space is calculated. A minimal spanning tree is then drawn between these points. The object of this tree is to connect points to each other that are most similar. The distance used for this connection is related to the Euclidean distance calculated above. A minimal spanning tree uses a line to do this, where similar points are connected next to each other on the line. The line is drawn in such a way as to result in the least overall length possible. The line cannot fold back on itself creating circuits. This means that no point can be passed through a second time. This resulting tree then has points connected to each other that are most similar. The starting point for creating such a tree is arbitrarily chosen.

A spanning tree can help to determine some criteria for "group" definition. Since the length of the line segments between points is a reflection of the similarity between the points, group definitions can use this as criteria for separations. Often clusters of cases are "pruned" from the tree on the basis of the length of the line segment between two points relative to the average lengths between all sets of points. If the length between two points is larger than some factor times the average length, it is ruled to be too long, and the cluster is "clipped" at this point. This distance criterion, therefore, defines the end of a category or group. The criterion used for this factor is user-defined. This can be altered to change the number of clusters or groups formed. There is no single "correct" number of clusters. The "closeness" accepted for a cluster has to be defined by the experimenter.

DRAWING OF THE TREE

TREE does not draw the minimal spanning tree. This can be done manually if the number of samples is reasonable. Results from TREE are given in Table II.4a for the summer air pollution data at the MC and NS stations. Patterns are listed by case number. Part A of the table gives a summary of the case number breakdown. The case numbers are listed that represent each station for each month during each year. Since

Minimal Spanning Tree for Summer Data

Part A. Summary of case numbers belonging to each sampling station for each month (i.e.: MC75JN contains all Maricopa County samples for June 1975)

Case Numbers	Samples	Case Numbers	Samples
1,2,3	MC75JN	31,32,33	NS75JN
4,5	MC75JL	34,35	NS75JL
6,7,8,9,10	MC75AG	36,37,38,39,40	NS75AG
11,12,13	MC76JN	41,42,43	NS76JN
14,15,16,17	MC76JL	44,45,46,47	NS76JL
18,19,20	MC76AG	48,49,50	NS76AG
21,22,23	MC77JN	51,52,53	NS77JN
24,25	MC77JL	54,55	NS77JL
26,27,28,29,30	MC77AG	56,57,58,59,60	NS77AG

Case numbers are used in Parts B and C of the Table to signify these samples.

Part B. Sample output from TREE (only nodes for first few steps in the drawing procedure are given)

```
NODE 1
       NODE 41   AT DISTANCE   .2173
       NODE  2   AT DISTANCE   .2298
       ....
NODE 2
       NODE  1   AT DISTANCE   .2298
       NODE  8   AT DISTANCE   .3500
       ....
NODE 6
       NODE  8   AT DISTANCE   .2112
       NODE 10   AT DISTANCE   .2667
       NODE  5   AT DISTANCE   .3399
       NODE 38   AT DISTANCE   .4191
       ....
NODE 8
       NODE  2   AT DISTANCE   .3500
       NODE  6   AT DISTANCE   .2112
       NODE  4   AT DISTANCE   .3124
       ....
NODE 41
       NODE 52   AT DISTANCE   .1710
       NODE 29   AT DISTANCE   .2148
       NODE  1   AT DISTANCE   .2173
       ....
```

(Continued overleaf)

TABLE II.4a *(Continued)*
Minimal Spanning Tree for Summer Data

TREE Drawing with MC and NS Cases

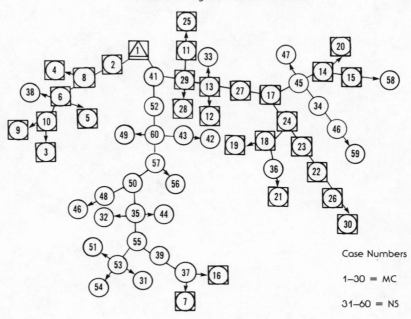

Case Numbers

1–30 = MC

31–60 = NS

there are two stations, three months, and three years, this gives a breakdown into 18 groups.

Part B of Table II.4a shows a sample output from TREE in ARTHUR. ARTHUR uses the word "node" to indicate the point on the tree where a case or pattern is present. Therefore, node 1 is the location on the tree of case number one. From Part A it can be seen that this represents a MC75JN sample. Only those cases necessary for the initial steps of the drawing are listed in Part B. The total printout would give a listing similar to these for all 60 cases.

From Part B, it can be seen that node 1 is connected on the tree to the samples with case numbers 41 and 2. Part A of the table shows that these are samples from NS76JN and MC75JN, respectively. Distances from case one for these two neighboring cases are also listed in the printout (Part B) with the NS76JN case being a distance of 0.2173 and the MC75JN one being 0.2298 (units of the distances are arbitrary but can be used for comparisons). Therefore, if the tree was to be drawn, the first step would be to put case 41 a total of 0.2173 units away from case 1 and case 2 at 0.2298 units away from case 1. If relative distances are to be kept significant, one must note also the distance between cases 41 and 2 to calculate the angle between these. This procedure leads to long time-consuming drawings. It is felt, however, that a tree drawn without the precise distances taken into consideration is almost as useful and much quicker.

Although actual distances drawn are then not relevant, neighboring cases are still similar and much information can still be gained. Taking this approach, the first step is to draw

where the triangle signifies at which point the analysis was begun, and the circle represents all others. The next step would be to find the neighbors for case 41. Continuing down the printout in Part B until this point is reached, it can be seen that the neighbors are case 1 (as expected) and cases 52 and 29. Similarly, case 2 has neighbors of cases 1 and 8. Therefore at this point, the TREE diagram would look like

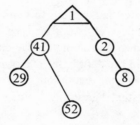

with distances and angles being insignificant. Often times a node may have more than two neighbors as with case 41, but this causes no problems. Another example of this can be seen by completing the next two steps of the drawing from node 8. This node has three neighbors, 2 (as expected) and nodes 6 and 4. Node 6, however, has three other neighbors. Each of these is placed as a separate branch out from node 6. Any number of branches can exist. This step would then look like

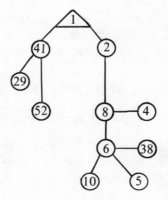

Again, distances and angles are not significant. Neighbors are then sought for nodes 29, 52, 10, 5, 38, and 4. Results show that both nodes 38 and 5 have no other neighbors

besides node 6. Arrows can be used to designate such places where branches end. This is done only to aid in the drawing process. The completed drawing is shown in Part C of Table II.4a. In this drawing, the case number is given inside of the circle. Squares are used to designate MC samples and circles for NS. These are used to help visualize the results obtained.

CLUSTER DEFINITION

The next step would be to define what criterion is to be used for "clusters." As was stated previously, there is no correct number of clusters to be obtained from this tree. The two extreme examples would be either to consider each case a cluster, or consider all cases as one single cluster. The optimal solution is generally somewhere in between these. TREE has the ability to give clusters if certain "pruning parameters" are chosen. The pruning parameter is essentially the factor times the average distance between cases that is chosen. If a distance then exceeds this value, the cluster is clipped at this point. This can prove to be quite useful, but often a few single points become their own clusters and all the rest are thrown into a single large cluster, which occurs when there exist a few truly unique points in the data base. The pruning parameter must be experimented with, to find optimal cluster definitions. It is also useful to use the tree for studying trends and the next step, hierarchical clustering, for actually defining the groups.

It must be noted that this tree was arbitrarily begun at case 1. Since distances are meaningless on this diagram, the two extreme sides are deceivingly far apart. In reality, this two-dimensional drawing may really fold back on itself so cases 58 and 9 may be much closer together than indicated (although they are not "neighbors," i.e., adjacent to each other in this Euclidean space). This is a problem encountered in all n-dimensional space manipulations. The computer can calculate exact distance values in n-dimensional space without error. However, relaying the information obtained into a two-dimensional space or some type of tabular form always results in some lost information. This is why it is critical never to rely on a single multidimensional statistical technique. Only through comparisons of the results obtained from many statistically unrelated programs can truly significant results be obtained.

MINIMAL SPANNING TREE RESULTS

When studying the tree in Table II.4a, there may be three initial questions to ask:

1. Do the samples from the same collection site cluster together?
2. Do the samples from the same month regardless of station site cluster together?
3. Do combinations of these two factors dictate similarities?

Usually the answer is not simple. If distinctions were so great that all samples followed one rule or the other, their characteristics were probably so different initially that they could have been picked out of the raw original data, making the usefulness of the multivariate techniques minimal. Often, only certain preliminary ideas of trends can be seen at this step. Optimal uses of these techniques are for problems such as this one where exact easy answers are not available. The data must be studied in a variety of ways to try to extract useful information from them.

TREE offers no simple answer. The top part of the diagram is dominated by MC samples (case numbers up to 30) with some mixing of NS. A thorough study of the results needs to be made at this point in the analysis. It would be very useful to perform a second unsupervised learning technique so results of the two can be compared. The second type will also be easier to use when trying to distinguish class membership.

STEP II.4b. HIERARCHICAL CLUSTERING

The next unsupervised learning technique performs a hierarchical clustering in Step II.4b. Three programs in these packages are available for this, HIER in ARTHUR, BMDP2M, and BMDPKM. There are many possible algorithms available for hierarchical clustering techniques. Distances between every pair of points must again be first calculated to define "similarity." This is identical to TREE. Hierarchical clustering differs from TREE in the display chosen (and therefore the errors incurred) for summarizing *n*-dimensional data into two dimensions. Of the three programs available, HIER in ARTHUR will be discussed first. BMDPKM and BMDP2M will be considered later.

HIER proceeds similarly to BMDP1M used in Step II.3d. Originally each point is considered as a single "cluster" (other possibilities will be discussed later). The next step is to find the shortest distance between any pair of points. These two points are then considered most similar and are combined into a single group. The distances from this group to all other points is then calculated using some type of "average point" to represent the center of the two points' locations. The second shortest distance is then found (whether it is between this group and a point or two other points). This pair is considered second most similar and is combined to form another group. This process is continued until the entire data set is included in a single group. See Step II.3d for more details of this process. A dendrogram is then drawn. An example for these data is given in Table II.4b. This printout was taken from HIER in ARTHUR. Only those cases which were members of categories MC or NS were again considered.

DESCRIPTION OF THE DENDROGRAM

In this table, the horizontal axis represents similarity measurements. The left value of 1.0 means the two patterns are identical. As you proceed to the right, similarities between points decrease. Vertical lines indicate that two clusters are being joined at that given similarity. HIER begins with every pattern being considered as an individual cluster. Only identical points would be joined at a similarity value of 1.0. As you proceed to the right, the similarity criterion for cluster membership is relaxed and therefore clusters may be joined. The process ends when the entire data set becomes a single cluster (somewhere before a similarity measure of zero).

TABLE II.4b
HIER Plot of MC and NS Cases

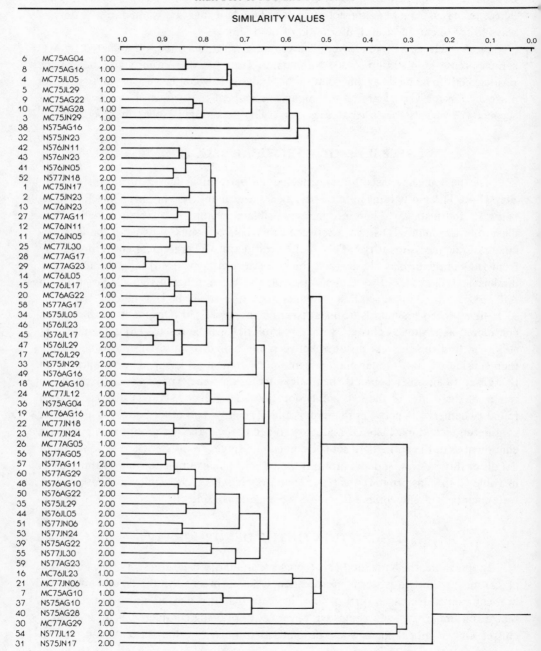

The cases or patterns are listed down the left vertical axis. Neighboring patterns indicate similar points. However, it must be remembered that projection errors still exist in this technique. In a dendrogram, each point (except for the first and last) must have only two neighbors. This may not be the case in n-dimensional space. But this criterion is necessary to give a two-dimensional idea of the n-dimensional similarities. The program TREE does a better job of illustrating the true near neighbor density for a point than HIER, but cannot show similarities or distances as well. As with TREE, HIER must begin at some point. The distance between the two ends may not be a true representation of what really exists. It is only through comparisons of the techniques that meaningful results can be obtained.

In Table II.4b, cases 28 and 29 (MC77AG17 and MC77AG23) are most similar, combining at a similarity value of 0.94. These are found approximately one-third of the way down the table. These are connected with a vertical line at this similarity value. As the acceptable value of similarity is decreased, more groups are formed. If the accepted value is lowered to 0.90, samples 34 and 46, and 45 and 47 are also combined. As similarity values are further decreased, two possible results exist. The first is shown above, where other pairs of points may then meet the acceptance criteria and be combined. A second possibility would be that an additional point (or cluster) is added to a cluster already containing two or more points. Both can be handled. The only difference comes in the definition of distance or similarity. The distance between two points uses an n-dimensional Euclidean distance. The distance between a point and a cluster must first define the location within the cluster from which the distance measurement is to be made. Many options exist and are discussed in Steps II.3d and II.4d which offer many choices for this decision.

Consider the first four patterns on the dendrogram in Table II.4b. Cases 6 and 8 are joined at a similarity of 0.84. Case 4 is joined to these to make a three-point cluster at 0.74. Case 5 is added at 0.73. There is no "correct" number of categories or points within a given category. The acceptable similarity value must be defined as the dendrogram is studied.

From Table II.4b, no well-defined category definition is seen. The top seven patterns all belong to class 1.00 (MC). There then occurs a group of class 2.00 followed by more 1.00's, etc. The data seem to be splitting into subgroups, mostly along pattern lines but not strictly adhering to these rules. Note that case 17 (MC76JL29) halfway down the table fits better in with a group of July NS samples than with MC samples. Also note that the last three points on the dendrogram, cases 30, 54, and 31, are not very similar to any other points, only joining at a similarity of near 0.30. This dendrogram clearly points in the same direction as those results from TREE. No simple answer to the categorization of data question exists. However, some subgroups of samples are quite well defined.

This dendrogram does contain a large amount of useful data. The next step in an analysis should be to compare critically these results with those obtained previously with TREE.

OTHER HIERARCHICAL CLUSTERING PROGRAMS

BMDP2M is another program similar to HIER. It also begins by considering each case as an individual cluster. Four distance measurements are available. Most commonly used is the Euclidean distance. This technique does not use iterative distance calculations (distances are not recalculated after each step). This makes BMDP2M much faster in computer time (HIER has been known to use large amounts of computer time) but results are not as good. For reasonably separable data, both techniques give similar results. However, for difficult cases, HIER uses more computer time but gives superior results. Also, the printout for BMDP2M is similar to that for BMDP1M (see Table II.3a), which we feel is more difficult to interpret than HIER (see Table II.4b).

The technique of considering each case as an individual cluster initially is called agglomerative. Another clustering method originally considers that all cases are contained in a single cluster. This is referred to as diverse clustering, and this is available in BMDPKM, a program added in 1979. This single cluster would be divided by some chosen criterion. BMDPKM uses a Euclidean distance between the cases and the center of the cluster. This first division can be performed in $2^{(n-1)} - 1$ ways. For a large value of n (number of cases), this quickly becomes prohibitively large. Oftentimes a figure of $n = 16$ has been used as a reasonable maximum number of cases for this method.

The process continues to divide the clusters until either (a) the requested number is reached or (b) each case represents a given cluster. Usually (a) is chosen. Cases are then iteratively relocated into the cluster whose center is closest to the case. An optimal solution is obtained. For reasonably small data sets, both agglomerative and diverse clustering should be tried, and results compared.

Distances between groups can also be defined in a variety of ways, and the methods for calculating the location of the "average point" for a group vary. These "average points" can often be weighted to reflect the number of cases in the group. Many of these options are available in the computer program CLUSTAN which will be discussed at the end of this section. If regular use of clustering techniques is a major part of your statistical manipulations, it may be worthwhile to familiarize yourself with this program as well. If not, the routines in ARTHUR and BMDP are quite adequate.

STEP II.4c. NONLINEAR MAPPING

The third unsupervised learning technique is nonlinear mapping. This is found in subroutine NLM of ARTHUR. Two-dimensional plots of the n-dimensional data are made, trying to minimize the interpattern distance errors. While HIER and TREE preserve local structure (patterns close to each other on the dendrogram and tree are truly similar while patterns far away may actually be closer together since these drawings may fold back on themselves), NLM preserves global structure. Small distance errors are much less important than large ones. The plotting of distance is defined as the interpoint Euclidean distance in the two-dimensional projection plane of the n-dimensional space. To find this becomes an iterative process since minimization of error employs changing the position of the points in the two-dimensional space until an optimal location for each point is found. The process can therefore also be expensive in computer time. For this reason, TREE is run first to get initial ideas of trends for the smallest amount of computer time.

An NLM plot from ARTHUR for MC and NS summer data is given in Table II.4c. ARTHUR allows three choices. Either the case number, station number, or case name can be plotted. In this table, the case number is plotted and those cases belonging to category MC are circled. Trends can be studied along station lines. Thirty iterations were performed to obtain this plot. The iteration limit can be user-defined.

No outstanding trends in the data are shown. Category 1 does occupy the top half of the figure, in general, while 2 is confined mainly to the lower left. Overlap between the two is seen (as was the case with HIER and TREE). Individual points of interest can be identified from this case number plot. Note that the point belonging to category NS in the upper left area is case number 38 which is identified (from HIER printout in Table II.4b) as NS75AG16. This plotted relatively close to group 1 samples in the HIER plot as well. Other individual points can be studied as desired.

TABLE II.4c
Plot of Case Numbers for MC and NSNLM Analysis

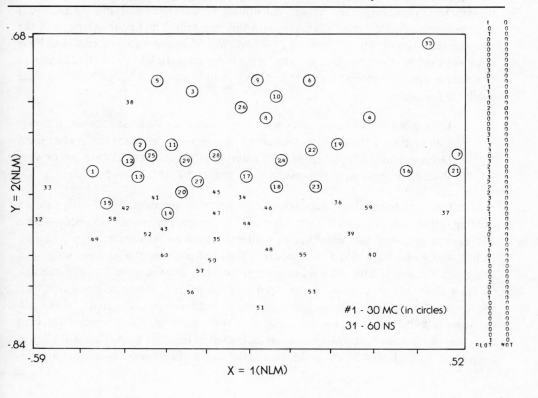

#1 - 30 MC (in circles)
31 - 60 NS

SUMMARY

The three unsupervised learning techniques given above are very complementary in their statistical approach. Usually, a comparison of these three gives very good unsupervised learning results. Much information is contained here. The computer is defining trends in the data with no *a priori* knowledge of the true categories that do exist (if these are indeed known). If category numbers are assigned, the computer does not take them into consideration in these three techniques. The key, therefore, lies with the experimenter, who must make sense of the results from above. The combination of the three techniques defines the trends, but it is left to the user to interpret these. Here lies the toughest part of the analysis. One must avoid overinterpretation while summarizing the results in a meaningful manner. Unfortunately, no rules for doing this can be given. It is only through practice in interpretation of these results that success in this endeavor can be obtained.

STEP II.4d. CLUSTAN

To supplement these unsupervised learning techniques to define group structures, CLUSTAN may be used (Step II.4d). This program is solely dedicated to unsupervised learning techniques. For general overall data analyses, Steps II.4a through II.4c should be sufficient. However, if the major goal utilizes unsupervised learning frequently, it may be worthwhile to become familiar with this program. The major extension of the program lies in its variety of "similarity" definitions and output display techniques. Four steps are used for most hierarchical clustering techniques. In CLUSTAN, a variety of procedures are available at each step. These include (with CLUSTAN subroutines available for use):

a. Data input and distance and correlation measurements (FILE, CORREL, DISTIN)
b. Calculation of the clustering scheme (HIERARCHY, MODE, DENSITY, and KDEND)
c. Decisions on the definition of the clusters present (DNDRITE, EUCLID, RELOCATE)
d. Plotting techniques of the results (SCATTER, PLINK, TREE)

FILE is used to input the data into the program as raw data. DISTIN will read an existing distance matrix into the file. This can be a correlation matrix (example—from SPSS) or cross comparison data that were directly observed. An example would be data from doctors who were asked to compare a group of patients qualitatively instead of using raw numerical data as could be obtained from a questionnaire. CORREL can calculate a similarity matrix for use as input into the other techniques, many of which require this type of input. Dissimilarity coefficients can also be calculated. CORREL can calculate a k-linkage list. Given a value of k from the user, this program will then calculate the k nearest neighbors to each individual in the data base. This is similar to Step II.5a. In CLUSTAN, the results are saved and utilized for other subroutines.

HIERARCHICAL TECHNIQUES IN CLUSTAN

After these preliminary calculations have been made, a clustering scheme is chosen. Here lies the heart of hierarchical clustering techniques. Many different methods can be used and a comparison of results from a variety of these can give great insight into the data structure.

HIERARCHY is a method similar to HIER in ARTHUR. Given n cases, each case is originally considered as its own cluster. Similarities between cases are calculated, and the two most similar are fused into a single cluster. A method then is needed to calculate the similarity between this new cluster and the other $(n - 1)$ cases. The next most similar pair of clusters is now fused. This could consist of either the two-point cluster and another case, or two one-point clusters (two individual cases). This process continues until a certain *a priori* defined number of clusters is reached or all points are joined in a single cluster.

The differences in the various techniques lie in the definitions of "similarity." Eight choices are available in subroutine HIERARCHY. These include the following:

a. SINGLE LINKAGE—Similarity is defined in terms of the closest pair of points, one of which belongs to each of the two groups under consideration. This type of method tends to find elongated clusters.

b. COMPLETE LINKAGE—Similarity is defined in terms of the furthest pair of points, one of which belongs to each of the two groups under consideration. This generally leads to tighter spherical clusters.

c. AVERAGE LINKAGE—The average of the similarity coefficients for all possible pairs of points, one of which belongs to each of the two groups under consideration, is calculated. This is used to define similarity between these two groups.

d. CENTROID—Mean vectors of the two groups under consideration are used. This distance between these is used to define their similarity.

e. MEDIAN—The distance from cluster A to a cluster formed from B and C is defined as the distance from the centroid of A to the midpoint of the line that joins the centroids of B and C.

f. WARD'S METHOD—An "error sum of squares" for each cluster is measured as the sum of the distances from each individual in the class to the centroid of the class. The two clusters are then combined, which minimizes the increase in this error.

g. LANCE WILLIAMS' FLEXIBLE BETA METHOD and

h. MCQUITTY'S SIMILARITY—both use a function relating the similarity between cluster A and the cluster formed from B and C as the average of the similarity of A to B and A to C.

Results from these methods in HIERARCHY can then be used to define and plot clusters.

OTHER CLUSTERING TECHNIQUES

Other clustering schemes can be calculated in MODE. The density of points around a given point is studied. This can be done with k nearest neighbors in CORREL. The average distance between this point and its k nearest neighbors is calculated. A small average distance reflects a high density in the data around that point. Individual points are then rank ordered according to these average distances. This ordering dictates the order for introduction of points to the cluster nuclei.

A threshhold value, R, is chosen to define clusters. If the distance from the second point to the first is greater than this threshhold, it is considered a new cluster of its own. If the value is less, point two is combined with point one to define a single cluster. This procedure continues until all points are considered.

One problem with MODE is the assumption of a constant threshhold value. With this restriction, definitions of both small and large clusters cannot be made simultaneously. Subroutine DENSITY monotonically increases R to define various spherical neighborhoods of points. Now both types of clusters can be studied.

KDEND uses a Jardine Sibson clustering technique. A data base with less than 60 patterns should be used. Similarities between all N objects are calculated. These are plotted as nodes on a tree graph. Now, all pairs of nodes are considered, and those with similarities above a certain threshhold value are connected. These subgraphs can then be connected, allowing for overlapping. This routine can become time consuming. After clusterings have been found RECODE can be used to attempt to optimize the solution of the problem. An initial classification into n clusters is necessary. Any of the previous techniques can be used to obtain this. Now consider a given case #1. The similarities of it to its parent cluster centroid and all other cluster centroids in the data base are calculated. If another cluster is closer, the case is moved to that one. The centroids are then recalculated (taking this move into consideration). Case #2 is checked in a similar manner and recalculations are made if necessary. This process continues until no more moves are necessary. This is considered the optimal solution for n clusters.

Next, the similarities between all pairs of clusters are computed and the two most similar are combined. The problem now consists of $(n - 1)$ clusters. The entire relocation process for each object is redone to obtain the optimal solution for $(n - 1)$ clusters. This continues until a minimal number (user specified) of clusters is obtained.

In comparison of these methods, RELOCATE and WARD'S usually result in tight clusters having a minimum within-cluster variance. MODE, SINGLE LINK, and KDEND usually define "natural clusters" that are often not as tight. It is through comparison of the various types that the best data can be obtained.

A definition of cluster cuttings can also be made with DNDRITE or EUCLID. These are similar to the cuts made in TREE in ARTHUR. DNDRITE optimized the error sum of squared or within group sum of similarities. Cuts are made to remove the longest connections within the cluster to minimize this value. With EUCLID, the Euclidean sum of squares is optimized.

Plots can then be made. SCATTER will create two-dimensional scatter plots relating distances between points similar to NLM in ARTHUR. SCATTER has available to it various forms of "connections" of similarity. PLINK and TREE in CLUSTAN are used to plot dendrograms of results similar to those obtained in HIER in ARTHUR. Note that the subroutine TREE in CLUSTAN is not similar to TREE in ARTHUR. Horizontal or vertical dendrograms or horizontal summary tables can be requested.

SUMMARY

In summary, CLUSTAN does offer a large variety of useful unsupervised learning techniques. The only problems envisioned in its use are in making meaningful interpretations of differences in the results obtained from the different techniques available. Also, implementation and documentation is not as good as with the other programs (although improvements have been recently made).

Step II.4 has therefore been used to study "natural" classes that exist in the data structure. These will be checked in the next step.

II.5. SUPERVISED LEARNING TECHNIQUES

INTRODUCTION

The next step in a statistical analysis usually entails the use of supervised learning techniques. If the data had an inherent category structure (that is, each was known to belong to a certain group), supervised learning could be used to study the structure of separation. Unlike Step II.2, and similar to Step II.4, the analysis is done multivariately. All variables are considered simultaneously instead of individually. Separability is studied in the data structure. If, however, known categories are not present in the original data, the programs discussed in Step II.4 can be used to find such "natural" groupings in the data, followed by Step II.5, which will help decide the degree of separation.

Another major use for supervised learning is to train the computer to recognize the category definitions in a training set. A figure of "trainability" or percent correct prediction for the known samples is calculated by the computer. Samples whose correct classifications are not known (or are withheld from the computer to check the predictability) can then be given as a test set. The computer will assign each of these samples to a class according to the criteria used to train the computer. The success of classification can be tested by this method.

In supervised learning, the techniques vary in the criteria used for defining class membership. Two major approaches are discussed here. Comparisons of results from these can be made.

These two techniques are statistically unrelated. This is important when determining "significance" of group separation. Clusters of data within a group can take on a variety of shapes. If the clusters consist of tightly grouped points with very small inter-group differences and large intragroup differences, the separation would be identical for most of the algorithms utilized. This, however, is generally not the case. The group may be elongated towards one axis or spread over a wide range of values. When this occurs, group definition is much more difficult. Also, the assignment of unknowns to a group becomes very dependent on a definition of group membership and how the point's "closeness" or "similarity" to a group is defined. In the techniques considered here, the first, KNN, is not concerned with group shape. It assigns membership according to "near neighbors." The second technique, discriminant analysis, is concerned with the absolute separability of the groups. Group membership is assigned according to location with respect to the plane separating the groups.

Other supervised learning techniques do exist. Also, the programs in Step II.4 can be used to check the validity of the group separation. Samples whose classifications are not known can then be added and group assignments for these predicted.

STEP II.5a. k-NEAREST NEIGHBORS

In the first Step, II.5a, subroutine KNN from ARTHUR is described. In KNN, k-nearest-neighbors technique, the distances in n-dimensional space between all pairs of points are calculated, similar to some of the procedures in Step II.4. For example, given sample A, the Euclidean n-dimensional space distances in the sample measurement space (with each measurement being an axis as before) between point A and all other sample points are measured. Group membership for point A is then defined as the group to which the k closest points to A belong. For instance, if $k = 1$, the group designation for A would be the same group of which the nearest point of A is a member. This completely ignores the shape of the groups and defines similarity to a group in terms of similarity to individual points within the group. Any value of k can be specified. If k is greater than unity, a "committee-vote" rule goes into effect. For example, if $k = 5$, the case is assigned as a member of the group most often represented by the point's five nearest neighbors. This procedure assumes that all points belong to some previously defined class. Therefore, outlying points are still classified as a member of the group whose point is closest to the outlier regardless of the real distance involved.

For the initial step of training the computer with KNN, a sample printout like the one in Table II.5a is obtained. For this analysis, the entire summer data base was used. In this table, only samples 5 through 7 are shown. Similar printouts are obtained for each sample in the set. The ones given here were chosen to illustrate a few points.

The first sample listed is case number 5, MC75JL29. The known category for this sample is category one (those samples from sampling station MC). Therefore, "TRUE CAT" is one.

Results are then listed across the page for one nearest neighbor and committee votes for 3, 4, 5, 6, 7, 8, 9 and 10 neighbors. Two-nearest-neighbor committee votes are not used. Since oftentimes the two would result in a split decision, it would be difficult

TABLE II.5a
k Nearest Neighbor Analysis (KNN)

Part A   Data Output

INDEX	NAME	TRUE CAT	1	3	4	5	6	7	8	9	10	INDEX	CAT	DISTANCE	INDEX	CAT	DISTANCE
														NUMBER OF NEAREST NEIGHBORS			
5	MC75JL29	1	1	1	1	1	1	1	1	1	1	6	1	3.399-01	8	1	3.472-01
												4	1	3.940-01	2	1	4.332-01
												32	2	4.430-01	38	2	4.499-01
												10	1	4.582-01	1	1	4.938-01
												25	1	4.993-01	11	1	5.302-01
6	MC75AG04	1	1	1	1	1	1	1	1	1	1	8	1	2.112-01	10	1	2.667-01
												5	1	3.399-01	4	1	3.739-01
												3	1	3.851-01	38	2	4.191-01
												9	1	4.289-01	24	1	4.488-01
												2	1	4.566-01	18	1	4.794-01
7	MC75AG10	1	2	2	2	2	2	1	2	1	2	37	2	3.317-01	40	2	4.387-01
												3	1	4.874-01	39	2	5.068-01
												24	1	5.161-01	8	1	5.409-01
												10	1	5.482-01	36	2	5.542-01
												18	1	5.544-01	48	2	5.730-01

Part B   Summary of Data from KNN Using k =1 Results

Correct Station– 46%		Correct Month– 66%		Broken Down by Station	
MC	37%	Jn	69%	MC	60%
MS	47%	Jl	53%	MS	67%
NS	67%	Ag	71%	NS	60%
SC	33%			SC	77%

to have a majority rule, and it is therefore eliminated. Results for committees of three through ten members are listed with each neighbor getting an equal vote. In the case of a tie, the category having the nearest neighbor wins.

On the far right side of the printout, the individual cases that are these nearest neighbors are listed. The single closest point to sample 5 is sample number 6, which belongs to category 1 (MC). The distance between these two points, 0.3399 units, was the shortest distance between sample 5 and any other point in the data set. The distance measurements are identical to those in TREE. The units are arbitrary, but may be used for internal comparisons.

Reading across, the second nearest point to sample 5 is number 8, a member of category 1. Note there that if a two-member committee vote was attempted, it would result in a prediction of category 1 membership since each of the two nearest neighbors belongs to that category.

The third nearest neighbor is sample number 4, a member of category 1, at a distance of 0.394. The three-member committee vote would also be category number 1.

This process continues up to ten members. Note that the first four neighbors are all members of category 1, the fifth and sixth belong to category 2, and the seventh through tenth to category 1. Therefore, this point could be considered as a rather well-defined member of category 1.

Sample number 6 (the second one listed in the table) is a member of category 1. The first five neighbors also belong to that category. The sixth is a member of category 2 (MS) and the seventh through tenth belong to category 1. Again, this sample is a well-defined category 1 member.

Sample number 7 (the third one listed in the table) is a member of category 1. The nearest neighbor is sample 37, a member of category 2 at a distance of 0.3317 units. The second nearest neighbor, sample 40 is also a member of category 2 and is at a distance of 0.4387 units away. The third nearest neighbor is sample number 3, a member of category 1. A three-member committee vote would still, however, predict category 2. Note the results of the committee votes. Considering 1, 3, 4, 5, or 6 member committees, the result obtained is that sample number 7 resembles class 2, not class 1, although it is known to be a member of class 1. Sample 7 is, therefore, not a well-defined member of that class when the KNN algorithm is used.

A similar argument continues for all samples in the summer data base. Summary results are shown in Part B of Table II.5a. These reflect the dismal results obtained for the data base from the single nearest neighbor technique. Overall success was only 46% with NS reflecting the highest success rate at 67%. These data are indicating that the classification definition that has been forced on them (which is separation by station location) is not reflected in the chemical data. The stations are not separable chemically. This was also seen in Step II.4 with the unsupervised learning techniques. Supervised learning further indicates which samples cannot be correctly classified using these group definitions. The procedure further indicates to which samples and groups each case is most similar.

A redefinition of classes was made, according to months: June (Jn), July (Jl), and August (Ag). Better results were obtained. These are shown in Table II.5a, Part B. June was correctly classified 69%, July 53%, and August 71%. For these, the correct month in SC was most often predicted with a success rate of 77%.

A further study was made with KNN, including only those samples from either MC or NS for the three summer months. The correct station was predicted in 83% of the cases, and the correct month 67% of the time.

This again reflects results similar to those obtained from unsupervised learning techniques. No easy classification system is reflected in the chemical data. However, insight into the problem is being gained.

TESTING SUPERVISED RESULTS

Assuming samples were also present with unknown classifications, the next step in the analysis would be to include these as a test set. After the computer had been trained with KNN, classification predictions for the test samples would be made. In this example, no test set existed. Since classification definitions were made by sampling sites, each sample included belonged to a known class (defined by its sample location). Therefore, no real unknowns existed in the data base. What could be done, however, would be to make samples from known stations the test set by deleting the sample site information from the data. The computer would then assign the sample to that station to which it was most similar with no *a priori* prejudices about its true classification. A check on the computer's ability to correctly classify it could then be made.

This type of unknown classification is much more useful in defining category definitions whose significance is directly related to the variables used as features. An example is a classification of rock types based on chemical and petrological data. Unknown rocks could then be classified as a given type based on their chemical data. The rock type classification would be directly determined by the data. In the air pollution example, the classification was defined either by location or date of sampling, and was not necessarily directly related to a sample type. This was seen in the results obtained. A more complex classification scheme was needed to relate chemical data to sample type.

STEP II.5b. DISCRIMINANT ANALYSIS

In Step II.5b, a discriminant analysis method is used as a supervised learning technique. In this process, the relationship between a known grouping of the data and a list of "discriminating" variables is studied. These variables are chosen according to their characteristics that differ between the groups. These variables are then linearly combined and weighted so that the groups are forced to be as statistically distinct as possible. This is done by choosing the linear combination of variates that maximizes the one-way ANOVA F test, which tests the equality of the means for the linear combinations. This method requires the assumption that the covariance matrix is the same for each group. The method is quite sensitive to this, although only the program SAS has a built-in check for it.

More than one linear combination of the discriminating variables may be necessary. The maximum number generated is either one less than the number of groups to be separated, or equal to the number of variables, if this number is less than the number of groups. The linear discriminant functions take the form

$$D_i = (d_i)_1 v_1 + (d_i)_2 v_2 + (d_i)_3 v_3 + \cdots + (d_i)_p v_p$$

where D_i is the discriminant function score for the ith function, $(d_i)_1$ is the weighing coefficient for variable v_1, and v_1 is the value for variable v_1.

These functions are designed to keep the discriminant score, D_i, for the ith function similar within groups and different between the groups.

A second application of discriminant analysis is used to classify patterns into various groups. The rule for division is sought that minimizes the total probability of misclassification. This method requires that the multivariate normal distribution assumption holds, and the equality of the covariance matrices assumptions. This rule leads to the determination of the probabilities for each case being in the various groups. Checks on the success of this training step can be made. Separate linear combinations, similar to the one given previously, are obtained that are representative of a given class. These indicate the probability of membership in that class. The case is "most similar" to that class with which it has the highest probability. A percent correct prediction for the training set can be made.

Discriminant analysis can then be used to classify unknown patterns. An allocation rule for each group is used to assign new members. The unknown is then classified as a member of that group to which it is most similar (or to which it has the highest probability of belonging).

Discriminant analyses are supervised learning techniques. In a similar manner to KNN, a training set is used to "train" the computer to recognize existing classes or categories. The test set consisting of patterns whose true classifications are not known can then be added to the problem and group membership for these predicted. However, the criteria for assignment is different between the two techniques. Similar to KNN, the group shape is not considered. But KNN never addresses the problem of separability but only predicts according to individual nearness. Discriminant analysis attempts total separability of the groups and uses rules to define groups as a whole.

Various types of discriminant programs are available and will be compared. These include DISCRIMINANT in SPSS, BMDP7M, and MULTI and PLANE in ARTHUR. Initially, the discussion will be limited to DISCRIMINANT.

A printout for a typical run from DISCRIMINANT is given in Table II.5b. The data base used was identical to that in KNN. Summer pollution data was divided into the four county stations giving 120 patterns in four classes. The goal of this step then is to try to "discriminate" between the four classes based on combining the pollution variables in some linear fashion to create "discriminating" functions.

Immediately the investigator is faced with a decision, similar to that encountered in REGRESSION in Step II.3f. Variables may all be entered into the analysis simultaneously (called the "direct" method) or alternatively they may be stepped into the equation individually. As with REGRESSION this second method is preferred. Often more variables have been analyzed than are necessary to define separation. Only the most useful of these really need to be considered, and subsequently, time can be saved by knowing which few are most important to measure on the unknowns (saving time and money on data-generating steps). The stepwise procedure in DISCRIMINANT has five stepwise criteria available for choice. With each, however, the first variable is chosen to be the one that singly best discriminates between the classes. The second chosen is the one that then is best able to improve the "discriminating" value of the function. This process continues either until all of the variables have been selected or until the remaining variables are of minimal use in improving the discriminating power of the function. At the end of each step, the variables already included in the equation are rechecked. Often a variable previously selected can lose its usefulness in the equation due to the fact that its discriminating power may be equally defined by a combination of variables added later to the equation. If this is the case, the variable is removed from the equation since its contribution is no longer significant.

TABLE II.5b
Discriminant Analysis of Summer Data Base

Part A

Step 1
VARIABLE ENTERED ON STEP NUMBER 1..   MN

				DEGREES OF FREEDOM		SIGNIFICANCE
WILKS' LAMBDA	.72910	APPROXIMATE F	10.65116	3	86.00	.000
RAO'S V	31.95337	CHANGE IN V	31.95337	3		.000

--------VARIABLES IN THE ANALYSIS-----------			----------VARIABLES NOT IN THE ANALYSIS--------			
VARIABLE	ENTRY CRITERION	F TO REMOVE	VARIABLE TOLERANCE		F TO ENTER	ENTRY CRITERION
MN	10.65116	10.65116	CU	.97838	6.16071	5.49563
			FE	.14073	.34019	10.43033
			PB	.89958	10.70290	7.32472
			PART	.21886	2.40320	4.92502
			ORG	.94118	7.10114	5.08599
			SULF	.97788	1.80583	1.35237
			NIT	.99894	.84713	.89141
			CL	.93222	9.40395	7.00306
			CO	.98466	4.56179	4.22659

Part B. Summary of the analysis.

SUMMARY TABLE

STEP NUMBER	VARIABLE ENTERED	F TO ENTER OR REMOVE	NUMBER INCLUDED	WILKS' LAMBDA	SIG.	RAO'S V	CHANGE IN RAO'S V	SIG. OF CHANGE
1	MN	10.65116	1	.72910	.000	31.95337	31.95337	.000
2	PB	10.70290	2	.52920	.000	70.13952	38.18615	.000
3	CU	5.94659	3	.43650	.000	90.16462	20.02510	.000
4	CO	7.24223	4	.34594	.000	117.59613	27.43151	.000
5	CL	6.82532	5	.27682	.000	148.85651	31.26038	.000
6	PART	3.98752	6	.24182	.000	168.88886	20.03235	.000
7	ORG	2.89533	7	.21813	.000	183.18089	14.29204	.003
8	FE	1.58031	8	.20578	.000	193.76915	10.58825	.014
9	NIT	1.17429	9	.19689	.000	199.74742	5.97827	.113

DISCRIMINANT FUNCTION	EIGENVALUE	RELATIVE PERCENTAGE	CANONICAL CORRELATION
1	1.37149	59.05	.760
2	.66444	28.61	.632
3	.28671	12.34	.472

Part C
CENTROIDS OF GROUPS IN REDUCED SPACE

	FUNCTION 1	FUNCTION 2	FUNCTION 3
GROUP 1 MAR	-.28820	-.06902	-.77925
GROUP 2 MESA	.19839	.93096	.20848
GROUP 3 NSCT	1.46756	-.68301	.10741
GROUP 4 SCOT	-.79011	-.47587	.41889

Part D

CLASSIFICATION FUNCTION COEFFICIENTS

	GROUP 1 MAR	GROUP 2 MESA	GROUP 3 NSCT	GROUP 4 SCOT
MN	-31.46111	-50.60878	10.57152	-62.18303
CU	-.72479	1.65619	-.31898	-.47972
FE	.31796	.33314	1.52173	-.92611
PB	1.44075	.62037	.49570	4.61962
PART	.04154	.06352	.02889	.09556
ORG	-.02136	-.47153	-.74471	-.31727
NIT	1.76167	1.36154	1.87725	1.32658
CL	.93077	.19183	-.15392	1.05430
CO	.00121	.00182	.00086	-.00086
CONSTANT	-6.89998	-5.88685	-8.66177	-8.26341

(Continued overleaf)

TABLE II.5b *(Continued)*
Discriminant Analysis of Summer Data Base

Part E

SUBFIL	SEQNUM	MISSING VALUES	ACTUAL GROUP	HIGHEST PROBABILITY GROUP	P(X/G)	P(G/X)	2ND HIGHEST GROUP	P(G/X)	DISCRIMINANT SCORES FUNC 1	FUNC 2	FUNC 3
MAR	3.	0	1	1	.998	.839	4	.086	-.659	.315	-1.547
MAR	4.	0	1	1	.984	.521	4	.361	-1.170	.509	-.571
MAR	5.	0	1	1	.900	.897	2	.085	.200	.409	-3.397
MAR	6.	0	1	1	.774	.550	4	.440	-1.824	-.057	-1.228
MESA	1.	0	2****	4	.994	.650	1	.160	-.021	-.770	.942
MESA	2.	0	2****	1	.888	.448	2	.274	.946	.335	-1.460
MESA	3.	0	2****	1	.624	.713	2	.169	.916	.490	-2.407
MESA	4.	0	2****	4	.993	.470	1	.246	.106	-.622	.599

Part F

Prediction Results

ACTUAL GROUP	NO. OF CASES	PREDICTED GROUP MEMBERSHIP GR. 1	GR. 2	GR. 3	GR. 4
GROUP 1 MAR	23.	17 73.9%	1. 4.3%	1. 4.3%	4. 17.4%
GROUP 2 MESA	26.	8. 30.8%	16. 61.5%	0. .0%	2. 7.7%
GROUP 3 NSCT	15.	1. 6.7%	1. 6.7%	13. 86.7%	0. .0%
GROUP 4 SCOT	26.	5. 19.2%	2. 7.7%	0. .0%	19. 73.1%

PERCENT OF 'GROUPED' CASES CORRECTLY CLASSIFIED: 72.22%

Choices for selection of variables in the stepwise manner can utilize a variety of criteria. In brief, these include the following:

a. Wilks' lambda—a measure of group discrimination taking into consideration the "average" value of each group;
b. Mahalonobis distance—maximizes the distance between the two closest pairs of groups;
c. Minresid—minimizes the residual variance between groups;
d. Rao's *V*—a generalized distance measure that results in the greatest overall separation of the groups;
e. Maxminf—maximizes the smallest *F* ratio between pairs of groups.

The stepwise selection of variables continues until the partial multivariate *F* statistic is smaller than a given value. This statistic measures the ability of a new variable to discriminate between groups when the information contained in the variables previously selected is removed from consideration. Therefore, if none of the remaining variables increases the discriminating power, the analysis is considered complete.

RESULTS FROM DISCRIMINANT ANALYSIS

Table II.5b shows a typical printout of a stepwise analysis. Only the first individual step is shown (see Part A of the table) followed by the Summary table (Part B). Both

Wilks' lambda and Rao's V are given. The first variable chosen in the analysis is MN. It gave a lambda value of 0.73 and a V of 31.9. A one-way ANOVA is used to test whether MN values are significantly different at the four sampling sites. The p value obtained is zero, indicating that this is indeed true. For the variables not yet in the equation, the F to enter value, a partial F statistic after removing the influence of MN, for PB is the highest at 10.70. It would therefore be the next variable entered into the equation. In the Summary table, it can be seen that PB was indeed chosen for step number two. The Wilks' lambda value decreased to 0.53, and the Rao's V increased to 70.1 (indicating an increased distance separating the groups). This process continues with variables being added or removed according to given criteria. Nine variables were chosen of the possible ten given. Only SULF was left out. Note that with each step, (a) the F to enter or remove decreases, (b) Wilks' lambda decreases, (c) Rao's V increases (although the change in it is less for each successive step), and (d) the significance value finally rises above zero in the last three steps, showing the last variables to be of less importance than the first ones.

A summary is then given of the three discriminant functions. These are listed in decreasing discriminability, as can be seen by their relative percentages. All three are useful, although the third accounts for only 12.3% of the information. The canonical correlation is also a measure of the relative ability to separate the groups.

The program will also list the set of standardized and unstandardized coefficients for the discriminating function. These are not shown here. The first one would contain the largest analysis of variance F statistic value. The standardized ones are based on standard scores (see Appendix IX) having a mean of zero and a standard deviation of one. This is similar to the output of scores in REGRESSION in Step II.3f. The standardized score indicates the relative importance of the variable to the function.

The unstandardized scores are necessary to find the actual discriminant functions. These values are the coefficients necessary to find the linear equation for the function. By multiplying these by the raw values of the associated variable and adding the constant, the discriminant score is obtained for the function. This has been shown for the centroids of each group in Part C. Note that function 1 separates NS (1.47) and SC ($-.79$) from the other two. Function two is useful for MS having a value of 0.93 which is dissimilar from the other three classes ($-.07$, $-.48$, and $-.68$). Function three separates MC ($-.78$) from the other groups (.21, .11, and .42). Realize that these implications are greatly oversimplified. All three functions contribute to the overall separability.

The next section of the table (Part D) lists the coefficients for the classification functions. The classification equations can be used on the raw scores for an unknown. The value of the classification for the unknown in each group is compared. The unknown is then classified as belonging to that group that results in the highest score. For instance, using the results of this analysis, the value of the classification for MC for group one would be found by determining

$$C_1 = -31.5(\text{MN}) - 0.72(\text{CU}) + 0.31(\text{FE}) + 1.44(\text{PB}) + 0.04(\text{PART})$$
$$- 0.02(\text{ORG}) + 1.76(\text{NIT}) + 0.93(\text{CL}) + 0.001(\text{CO}) - 6.90$$

Values for C_2 (MS), C_3 (NS), and C_4 (SC) would be calculated similarly. A comparison of the four C values would determine the class prediction for the unknown.

In Part E of the table is listed a partial output from a case-by-case listing of the classification results for the training set. Classification scores are converted to probabilities of group membership, and assigned to the class with the highest value. *A priori* probabilities can be used for group membership if known (although not used in this example). For each case, a variety of results are printed. Note for MAR case #3 (a true member of group 1-MC), the group with highest probability of membership, $P(G/X)$, is group 1 at 0.839. The next most probable group is 4 with $P(G/X) = 0.086$. The discriminant scores for the three functions are listed next to this. Another statistic, $P(X/G)$, is shown. This indicates the probability that the known members of this group are at least as far from the group definition as is the case under consideration. Therefore, 99.8% of the MC samples are at least this far from a defined group "center." This aids in identifying those points which may not be a member of any of the groups defined in the analysis. The one chosen is only the closest of those choices given.

As with KNN, therefore, each sample in the training set can be checked for its classification against the rest of the group. With KNN this is done with respect to the other points most similar to the one in question. With DISCRIMINANT it is with respect to a group centroid or average.

A summary of the predicting ability for DISCRIMINANT is seen at the very bottom of Table II.5b, Part F. The overall correct classification is 72.2% (compared to 46% with KNN). Note that NS is correctly classified 86.7% of the time (in KNN 67%) with MS being least at 61.5% (in KNN 47%). This does not signify that DISCRIMINANT is a superior algorithm to KNN. Since they are different procedures, one or the other will usually give better results depending upon group structure, shape, and membership. The specific problem will dictate this. Therefore, both should be used and misclassifications compared.

It should also be noted that SPSS values for this classification are overoptimistic, since the same cases are used for determining the classification rule as are used to check its validity. BMDP has methods for minimizing this bias, including a jackknife method that leaves some "known" cases out of the classification procedure to use later during this check.

A variety of plotting options are available. A plot of the first two discriminant functions can be made with 1, 2, 3, 4, signifying MC, MS, NS, and SC, respectively. Group centroids are shown on such a plot. Each group can also be plotted on an individual graph. Overlaps of individual points within the groups can also be studied.

OTHER DISCRIMINANT PROGRAMS

BMDP7M is very similar to DISCRIMINANT. Stepwise selection utilizing forward and backward steps is available. Very similar output and results are present. Neither has any really marked advantages over the other one in the authors' opinion, although we feel the printout in DISCRIMINANT may be a little easier to read.

In ARTHUR, two types of discriminant analyses exist. In PLANE, used mostly to separate only two classes of data, a hyperplane in n-dimensional space is found that can separate these two classes. This hyperplane is a linear discriminant function as was previously discussed. With PLANE, a random set of weight vectors (similar to coefficients in DISCRIMINANT) is used. During the training, the weight vector is then used to analyze the data. Adjustments are continually made until all patterns in the training set are correctly classified (or a maximum number of attempts have been made). For problems concerned with more than two classes, this algorithm considers all possible pairs of classes. As the number of groups increases, however, this can become an oversized problem.

In MULTI, a similar problem is addressed, but for data having more than two categories, the training step creates $(n - 1)$ hyperplanes, where n is the number of categories. Like the discriminant functions, one is needed to separate each group from all other groups. Therefore, if the data contain five groups, four hyperplanes are created, with the first one separating the first group from all other groups, etc.

Table II.5b "PLANE" shows a partial printout from PLANE. All data were used, and the groups were considered pairwise until all combinations were studied. This table shows results for classes 1 (MC) and 2 (MS). After 501 cycles (user defined), the two classes were only 78.33% separable. The best weighting vector found is listed. The values listed are the coefficients for the eleven variables in the data base. The results are then listed for each individual case. The discriminant score decision surface is set at zero. All MC samples should calculate negative scores and MS positive scores for correct prediction. Also, the magnitude of the score reflects the ease (or difficulty) in the prediction for that case.

For instance, note that for the first MC sample, MC75JN17, the discriminant

TABLE II.5b

PLANE

Results of Class 1 and 2 Separation from PLANE

******************************** CLASSES 1 and 2 ***************************************

MAJOR PASS	NO. OF MINOR PASSES	NUMBER RIGHT	PERCENT RIGHT
1	30	40	66.67
2	56	44	73.33
3	36	41	68.33
4	60	31	51.67
5	36	47	78.33

etc.

PLANE WAS UNABLE TO SEPARATE CLASSES 1 AND 2 IN 501 CYCLES.
THE BEST WEIGHTING VECTOR IT COULD FORCE ON YOUR CRUMMY DATA WAS
-1.13-01 9.16-01 -7.66-02 9.82-02 8.94-01 -6.22-01 -5.56-01 -1.88-01 -5.82-01
-2.46-01 4.99-02

TRAINING SET RESULTS USING THE BEST WEIGHT VECTOR

NAME	CLASS	DISCRIMINANT	NAME	CLASS	DISCRIMINANT
MC75JN17	1.	1.2511-01	MS75JN17	2.	1.3570-01
MC75JN23	1.	8.5878-02	MS75JN23	2.	2.8846-02
MC75JN29	1.	-5.6286-02	MS75JN29	2.	-2.3850-02
MC75JL05	1.	-1.4383-02	MS75JL05	2.	6.1513-02
MC75JL29	1.	-2.1807-02	MS75JL29	2.	4.4594-01
MC75AG04	1.	-1.5838-01	MS75AG04	2.	7.2034-03

score is 0.125. It is positive, indicating an incorrect prediction of MS. The second case, MC75JN23, scores 0.086, also incorrectly predicted, but closer to the zero discriminating division surface than the first. The third MC sample, MC75JN29, gave a score of $-.056$ using this discriminant vector, putting it correctly on the MC side of zero. The fourth through sixth MC samples were also correctly predicted as belonging to MC. This analysis continues for all samples. Notice that the MS samples are listed on the right. The MC samples caused more problems (ten gave incorrect positive values) than NS samples (three incorrectly gave negative values). The next step would be to adjust the discriminating function to try to improve the MC predictions.

Note that as the adjustment is done (see the trends of "PERCENT RIGHT" as the number of passes increases) the change may help or hurt the prediction ability. As an alteration in the discriminant function is made, it is decided (by the percent correct) whether it was a successful one or not. If so, it is retained. If not, it is removed and another attempt is made.

The number of cycles (501) is a default parameter for the program. This may be user defined. It is useful to increase the number for more difficult separations, remembering that the increase directly results in increased computer time necessary for the analysis.

The printout for PLANE will then compare classes 1 and 3 (MC and NS), 1 and 4 (MC and SC), and all such pairs of classes. For this pollution data, final prediction results from PLANE were given as

MC	vs. MS	80%
MC	vs. NS	97%
MC	vs. SC	70%
MS	vs. NS	98%
MS	vs. SC	85%
NS	vs. SC	100%

Note that NS is quite distinct from the other groups, causing a minimal amount of trouble. MC and SC are most similar. This was also seen with DISCRIMINANT.

MULTI does a multiple classification procedure, attempting to separate all classes simultaneously (not pairwise as with PLANE). This is similar to DISCRIMINANT. MULTI defaults to 2000 iterations in this attempt, although the user can again change this value. A partial printout for the pollution data from MULTI is given in Table II.5b "MULTI." The manipulations are similar to PLANE. The difference lies in the desire in MULTI for total separation and in PLANE for pairwise. After 2001 cycles, the data were only 61.7% separated, which is quite similar to the final results of DISCRIMINANT (71.4%). The weight vectors are listed for the four classes (each number representing the coefficient for that variable in the data set) and each case is then listed with its prediction result. In this case, the sample is assigned as a member of that category that gives a value closest to zero. In the example in the table, case 1, MC75JN17, is predicted as category 2 (MS) with a value of $-.305$. The second closest category of it is SC, number four, with a value of $-.361$. The true category, MC, gives a result of

TABLE II.5b

MULTI

Results of Data Separation from MULTI

MAJOR PASS	NO. OF MINOR PASSES	NUMBER RIGHT	PERCENT RIGHT
1	33	49	40.83
2	34	35	29.17
3	29	38	31.67
etc.			

AFTER 2001 CYCLES MULTI COULDN-T SEPARATE YOUR DATA. THE BEST WEIGHT VECTOR(S) IT COULD RAM THROUGH:

```
1  -1.77-01   5.65-01  -4.85-02  -4.23-01   2.51-01   3.23-01   6.56-01  -1.33-01   1.89-01  -4.46-02  -3.34-01
2  -1.75-01   6.93-01  -1.55-01  -3.97-01   6.57-01   1.47-01   6.74-01  -1.61-01   1.03-01   2.03-03  -2.78-01
3   1.69 00  -9.04-02   1.08 00  -7.56-01  -1.13 00  -1.22-01   7.12-01   1.74-01   1.27-01  -4.81-01  -3.70-01
4  -1.39 00   6.02-01   1.56-01  -7.05-02   1.06 00  -1.08-02   6.81-01  -5.54-02   1.86-01  -3.16-01  -3.13-01
```

NAME	CATEGORY TRUE CALC	CLASS	WEIGHT	CLASS	WEIGHT	CLASS	WEIGHT	CLASS	WEIGHT
MC75JN17	1 2	1	-3.71-01	2	-3.05-01	3	-4.14-01	4	-3.67-01
MC75JN23	1 2	1	-4.00-01	2	-3.59-01	3	-4.69-01	4	-3.88-01
MC75JN29	1 2	1	-3.34-01	2	-3.28-01	3	-5.42-01	4	-3.69-01
etc.									

—.371, and category 3 (NS) a value of —.414. This printout continues for each individual case.

COMPARISON OF DISCRIMINANT PROGRAMS

In summary, a variety of discriminant analysis algorithms are available. PLANE does only pairwise separations. With any problem containing more than four categories the number of pairs becomes large and summaries of result interpretations become difficult. It is useful in those cases when one might want to study subsets of categories that have special interests.

MULTI, DISCRIMINANT, and BMDP7M give similar results. MULTI cannot do stepwise variable inclusion. Final discriminant vectors therefore include all variables, even those that may be of minimal help in the separation. No plotting or statistical significance tools are available in MULTI. But results are similar and individual cases are listed to find the problem cases in the separations. Therefore, the authors feel that all three of these programs are approximately equally useful. DISCRIMINANT in SPSS would be the slight favorite, however, if all three were available.

II.6. VARIABLE REDUCTION

INTRODUCTION

There are many methods for data reduction manipulations that can be performed. Data containing two measurements can be reduced to a simple two-dimensional plot. This helps to visualize the trends contained within the data. Also data-reducing statistics such as means, standard deviations, and variances are helpful in data interpretation.

Histograms may be plotted. Each of these techniques is useful as a "summary tool" for raw data.

If three variables are considered, various graphic representations can still be utilized. Pins of various heights or different symbols or colors can be used on an ordinary two-dimensional plot to represent the third. Also, more than one two-dimensional plot may be made at various values of the third dimension. These can then be superimposed to create a three-dimensional image.

These types of display become much too cumbersome for greater than three-dimensional space. For n-dimensional space $\frac{1}{2}(n)(n - 1)$ plots would be necessary to display all of the data. This number quickly becomes unwieldly. Therefore, some type of reduction in the dimensionality of the problem is soon needed.

Many of these have been encountered in previous sections. The goals in these sections may have been more towards either supervised or unsupervised learning techniques, but all problems encountered in n-dimensional space, as pattern recognition problems are usually described, must somehow give results of the analysis on paper. Since paper is inherently a two-dimensional space, some type of dimensional reduction is necessary. All methods used for this purpose attempt to provide optimal two-dimensional projections according to some rule which defines what is to be preserved.

TOOLS FROM PREVIOUS STEPS

REGRESSION (Step II.3f) uses a form of data reduction. A single variable (the dependent) is plotted against a linear combination of chosen variables (the independent ones). This allows more than two variables to be included in the analysis with the resulting plot still being a two-dimensional figure. However, restraints are necessary. The relationship between the variables is not known *a priori,* since that is the information being sought. Their relationship then has to be restrained to some mathematical model such as linearity.

NLM (nonlinear mapping) is certainly a data reduction technique but was used in Step II.4c as a supervised learning technique. A representation of n-dimensional space is made into two dimensions, minimizing a distance error function. Again, a summary, two-dimensional plot is possible.

DISCRIMINANT analysis (as well as REGRESSION) is used as a method for stepwise selection of variables in order to utilize only those that have real meaning for the problem. The number of variables is reduced to the minimum number that still retains the information. This then reduces the n-dimensional space to m dimensions, representing only those m most important variables.

In this step, three additional data reduction techniques will be discussed. These are by no means the only ones available but will be used to demonstrate a few complementary possibilities that do exist.

STEP II.6a. SELECTION OF VARIABLES

The first technique discussed, Step II.6a, is the use of SELECT in ARTHUR followed by VARVAR. Given a certain grouping of data into categories, subroutine SELECT will choose the single variable that can best describe the separability of the groups. A weighting value is assigned to represent its ability to successfully separate the categories. A second variable is then chosen that describes the largest portion of the remaining variance. This is also assigned a weight representing its success in separation. This process continues to choose variables until all variables are selected or until the last selected value is below a certain criterion of usefulness.

This procedure is similar to choosing variables in DISCRIMINANT. In SELECT, however, the variables are not then combined into a linear equation. They are listed singularly in the order of importance of separation. What is commonly done is to then plot the two variables having the largest weights. This would represent the best simple two-dimensional variable plot possible to show group separation without using combinations of the variables. Therefore, the most important variables for this purpose can be recognized.

Table II.6a gives the results for a SELECT analysis done using the original data base

TABLE II.6a
Results from SELECT/VARVAR of Total Data

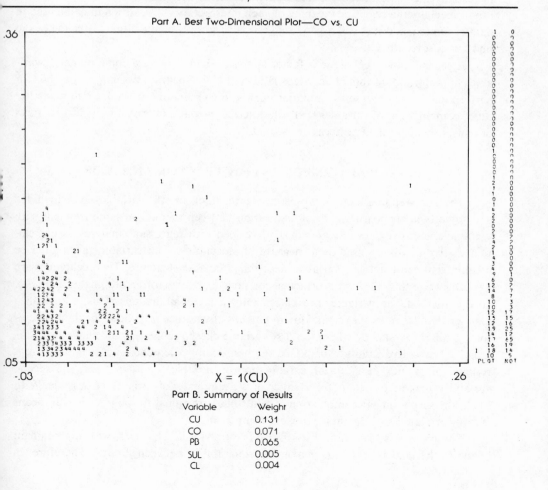

Part A. Best Two-Dimensional Plot—CO vs. CU

X = 1(CU)

Part B. Summary of Results

Variable	Weight
CU	0.131
CO	0.071
PB	0.065
SUL	0.005
CL	0.004

consisting of 456 patterns divided into the four classes. The single most useful variable was determined to be CU (with a weight of 0.131) followed by CO (weighted at 0.071). Only seven of the variables were above a specified tolerance level. Therefore, if the desire was to reduce the data base to contain only two variables while retaining as much separability as possible, the choice would be CU and CO. To visualize their usefulness, a plot using VARVAR in ARTHUR can be made. Either case number, case name, or category number can be plotted in this subroutine. With 456 points, only the category plot would be reasonable to study trends. This is shown in Table II.6a with CO on the y axis and CU on the x axis. The beginning indications of separability can be seen. Group 1 (MC) is starting to separate out from the other three categories. However, there is not total separation by any means.

To obtain a more successful separation, either more variables, or some combination of variables must be studied. There are advantages (and disadvantages) to each. As can be seen from this step, simple variable plots are often not enough to define separability, especially in data problems such as the one described in this example. As the data structure begins to utilize combinations and mathematical equations of variables, separability usually increases at the expense of a much more difficult interpretation of results (and especially the scientific meaning of such interpretations). Caution must be used in data reduction, and it is still a good suggestion to study a combination of techniques to find the best results.

A second major technique is factor analysis which utilizes a combination of variables to perform data reduction. Steps II.6b and II.6c illustrate two major methods for performing factor analyses. The major goal is to combine the variables into a smaller set of combinations of variables that will describe the data. Often underlying structures or causes for these data patterns are sought.

PRINCIPAL COMPONENT AND FACTOR ANALYSIS

The first step in a factor analysis problem is to choose the variables to be included and some type of measurement of association between them. The same problem with variable choice exists as was previously described with regression analyses. A correlation matrix is usually used as a measure of association. The variables must then be combined to construct new variables describing the data structure. Two techniques are available to do this. The first is principal component analysis (often designated as PCA). In this method, the variables are placed into linear combinations. This is done in a manner that accounts for the maximum amount of variance present in the data. Therefore, similarly to SELECT where the first variable chosen represents the single best variable to define separability, with PCA, the first combination chosen is that which represents the single "best" linear combination of variables. With SELECT, the second variable is chosen to account for as much of the remainder of variance as possible. With PCA, the second linear combination is said to be orthogonal to the first, which means it is uncorrelated with the first principal component.

The process is also similar to discriminant analysis. However, with discriminant analysis, the goal is to obtain maximum separability between groups. Therefore, in

PCA, a set of variables $x_1, x_2 \ldots x_p$ is transformed into a new set of variables $y_1, y_2 \ldots y_p$. Each y is made as a linear combination of the x's. This takes the form

$$y_1 = a_1 x_1 + a_2 x_2 + \cdots + a_p x_p$$

The first new variable, y_1, is designed to have the greatest variance of all possible combinations of x's. The second, y_2, must be uncorrelated with y_1 and contain the greatest amount of remaining variance. This process involves no assumptions about the relationships between the variables and usually no tests of "significance" are made. Few tests exist to decide if some principal components can be ignored, and usually this is left as a matter of judgment. The first few y's generated may reflect physical features of the original observations but the mathematic models do not assume that they do. This is left to the investigator. Nonlinear methods may also be used but extreme care must be exercised. Meaning may be "shown" in truly meaningless data.

STEP II.6b. PRINCIPAL COMPONENT ANALYSIS

PCA analyses can be found in ARTHUR in subroutine KARLOV (Karhunen–Loeve transformations) or as an option in FACTOR of SPSS. In Step II.6b KARLOV will be explained. However, if other factor analyses besides PCA are to be used, it may be easier to use FACTOR since both are obtainable in a single computer run. Methods for implementation for FACTOR will be given in Step II.6c.

Eigenvectors are extracted from the correlation matrix with KARLOV, and ordered from highest to lowest in terms of their eigenvalues. The percent of variance preserved in each can be calculated. This continues until either all the variance is explained or until the number of new "variables" (really linear combinations of the old ones) equals the number of old variables. Now, however, the maximum amount of variance is contained in the first few eigenvectors. A plot of the first two components for each data case can be obtained. Since these are linear combinations of the original variables, it is expected that this plot should contain more information about the data structure than any simple two-variable plot such as that obtained with SCATTERGRAM. It may be difficult, however, to assign scientific meaning to the linear combination.

Table II.6b shows results from a KARLOV analysis. Only the first four eigenvectors are listed. The data set contained the ten pollution variables for the 456 samples which were divided into their four sampling stations. Part A of the table summarizes the results. The first eigenvector is listed on the first line. Ten values are given, one for each variable. To obtain a measure of the amount of information preserved with this vector, the calculation becomes

$$(-.253)^2 + (-.168)^2 + (-.296)^2 + \cdots = 3.27$$

This value is then listed as the eigenvalue for the vector. Since the data were normalized before use, and each variable was adjusted to have a mean of zero and a variance of

TABLE II.6b

Karhunen Loeve (KARLOV) Plot of Total Data

```
ARTHUR, VER 8-14-76  **KARL**
ARTHUR DATA--  MC AND MS PLUS NS AND SC

I-M KARLOV (CALL ME BORIS).  I SMASH YOUR DATA FLAT.

NIN    11 CREATED IN SCAL NPAT 456 NTEST  0 NVAR  10 NCAT  4
NOUT   14

NPNT    0
NPCH    0

EIGENVALUE  INFO PRESERVED ****************************************EIGENVECTORS****************************************
            EACH  TOTAL

1  3.270+00   32.7  32.7  -2.525-01  -1.677-01  -2.964-01  -4.338-01  -3.727-01  -4.153-01  -2.801-01  -3.106-01
                          -3.603-01

2  2.174+00   21.7  54.4   4.750-01  -2.746-01   4.594-01  -2.920-01   4.269-01  -3.068-01   6.018-02  -4.228-03
                          -3.479-01

3  1.181+00   11.8  66.3   1.064-01   5.005-01  -1.839-01  -1.516-01  -2.364-03  -1.471-01   1.449-01   1.547-01
                          -3.018-01

4  8.420-01    8.4  74.7   2.209-02  -3.470-01   6.421-02   1.149-01   5.836-02   2.543-01  -5.616-02   5.247-01  -7.137-01
                          7.620-02
```

TABLE II.6b *(Continued)*
KARLOV Plot of the First Two Eigenvectors

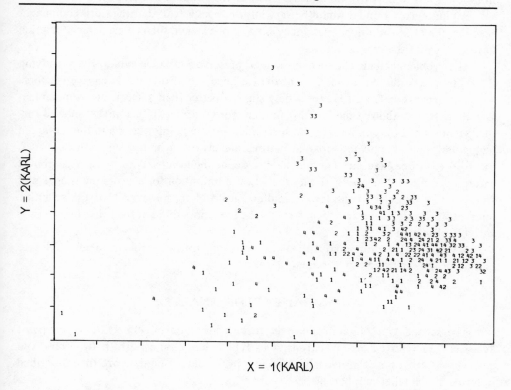

X = 1(KARL)

one, the total variance contained in the ten variables would be ten. Since the first eigenvector preserved 3.27 units of this variance, the percentage of the information preserved is (3.27/10) × 100 or 32.7%.

Similarly, the second eigenvector combination contained 21.7% more of the total variance. This is contained in that part not explained by the first. Therefore, the first two components would account for 54.4% of the total variance.

The third combination picks up 11.8% more of the total variance. Note how the largest amount is stacked in the first few eigenvectors. Ten vectors in all would be created. Only the first five are listed here. Note that these contain 82.8% of the information. Therefore, the last five (containing a total of only 17.2%) are not very significant. This then can be used to define a dimensional reduction for the data, retaining only those first few axes containing the largest portion of the variance.

A plot of the first two KARLOV axes is presented in Table II.6b, Part B. In ARTHUR, this is done, again, through subroutine VARVAR. The plotting choices are case number, number, or group. Assuming that the interest lies in group separation, the group number is plotted to study the success. This is shown in Table II.6b, Part B. The symbols 1, 2, 3, and 4 represent MC, MS, NS, and SC, respectively.

It can be seen that total separation is not obtained. Group #3 (NS) is trending, as a group, from upper left to lower right. Groups #1 (MC) and #4 (SC) are trending from lower left to upper right, and #2 (MS) is somewhat scattered in the middle.

A plot of axes 1 and 3 would be very informative. Recall that the third component accounts for 11.8% additional variance. The two plots could then be used in conjunction to study further the separation.

The problem can also be seen in the total preserved variance values. Even studying these three axes, there remains 34% of the variance assigned to the later eigenvectors.

The separation from KARLOV is only slightly better than SELECT, but complementary to it. SELECT showed the initial definitions for group 1 (MC). KARLOV showed further separations of group 3 (NS). Both routines lost much information in the reduction techniques. For complicated group structures such as that which exists in this data set, two (or even three) axes are not adequate to define the groups. Also KARLOV combined variables so more than two variables (as was the limitation for a two-dimensional plot from SELECT) could be considered simultaneously. Note, however, that the scientific interpretation of eigenvectors from KARLOV is much more difficult than that for the two-variable plot from SELECT.

If variables are eliminated with SELECT, it is sometimes possible to achieve a clearer picture of the separation by rerunning KARLOV.

CLASSICAL FACTOR ANALYSIS

The second type of factor analysis, to be considered in Step II.6c, is a type of "classical" or inferred factor analysis. This type is based on a faith that the observed correlations are a result of underlying factors in the data. The data are then described in terms of a smaller number of these types of factors.

In this procedure, each observed variable is divided into two parts. These are called the common and the unique. The common part of the variable is that part due to one (or more) of these underlying factors that describe the data base. This part will contribute to the variable's relationships to other variables in the data set. The second part, the unique, is governed by anomalies in the data or "background." An example will show this more clearly.

Consider for a minute the air pollution studies used as an example in this book. Measurements were made on ten variables, MN, CU, FE, PB, PART, ORG, SUL, NIT, CL, and CO. Most people who would collect such data probably have worked with pollution sources enough to begin at least to formulate sources for such pollution problems. It is known that automobiles contribute lead, organics, and particulates into the air. Copper smelters may contribute copper, particulates, and sulfates. Desert dust could add particulates, copper, manganese, and iron. Although these are very preliminary thoughts before the factor analysis problem is undertaken, a few points are to be made. First, note that we are assuming that certain "factors" are important in creating the data (dust, cars, and smelters may be three such factors). Each factor contributes to a number of variables that were measured. Therefore, one would expect to find some type of intervariable relationship between lead, organics, and particulates. This portion of the analyses for these three variables is referred to as the common part due to the car factor. Particulates should also show a correlation with copper, manganese, and iron, contributed by the dust factor. Similarly, a correlation between copper, particulates, and sulfates may be present. If these were the only three causes of the pollution, one would expect to be able to describe the original ten variables in terms of these three factors, each of which has some underlying meaning to the problem. This is rarely the case. Usually the major causes or factors can be determined. However, in almost all scientific data, there are present other minor factors that may not be distinguished as a factor, but result in slight increases in the variable measurement. This type of background increase due to minor causes is called the unique portion of the variable, not attributable to a single factor. This unique portion does not contribute any significant correlation between variables.

Realize that the assumptions of factor analysis require the underlying patterns to exist. Factor analysis, however, can in no way tell the investigator the causes for these patterns. The computer program only finds relationships. The investigator must then add the scientific meaning to such results, being careful not to overinterpret (or underinterpret) the computer results.

DATA CONSIDERATIONS

The mathematical assumptions and procedures behind factor analysis techniques are quite complex. These will be minimized here in order to concentrate on result interpretation. Each original variable, $V_1, V_2 \ldots V_j$ is divided into its contributions from each factor in the problem $F_1, F_2 \ldots F_i$ and a unique portion not attributable to a factor U_j. Therefore each variable can be written as a linear combination of its contribution from each factor as

$$V_1 = a_{11}F_1 + a_{12}F_2 + a_{13}F_3 + \cdots + a_{1i}F_i + d_1U_1$$
$$V_2 = a_{21}F_1 + a_{22}F_2 + a_{23}F_3 + \cdots + a_{2i}F_i + d_2U_2$$

The unique factor, U, must be uncorrelated with the common factors and all the other unique factors for the other variables. This assures that if any correlation between two variables is found, it will be related to some common factor(s) and not the unique ones. Therefore, when a minimal number of factors is extracted (each consisting of a set of correlated variables) the only thing left would be the uncorrelated "background."

FACTOR CHOICES

Before a typical output can be analyzed, a choice of the factor analysis method to be employed must be made. Five choices are available in FACTOR in SPSS. These include

  a. PA1—principal factoring without iteration;
  b. PA2—principal factoring with iteration;
  c. RAO—RAO's canonical factoring;
  d. ALPHA—alpha factoring;
  e. IMAGE—image factoring.

One of the most difficult steps in implementing factor analysis techniques lies in choosing among the options available. Although five methods are available, only two are yet commonly used. These will be compared here, leaving the others to either the SPSS manual or other factor analysis textbooks.

PA1 will extract principal components as in Step II.6b which are exact mathematical eigenvector transformations. This does not require any assumptions about the underlying structure of the variables. The goal is to describe the variance in the data in as few eigenvectors as possible. This choice can, however, be changed into an inferred method instead of a defined method. A value is assigned to each variable (from zero to one) that is used to represent the portion of variance of each variable that is to be assigned to "commons" (leaving the rest for the "unique"). No longer are exact transformations of the original data possible. This method is often called principal factor analysis (to distinguish from principal component analysis). The most difficult step lies in choosing this value (called the communality) for each variable.

PA2 is a method of principal factor analysis that chooses the values or communalities to be used. These are first estimated by the squared multiple correlation between a given variable and the rest of the variables in the set. This will, then, estimate the portion of that variable due to "common" or related to the other variables. An iteration process proceeds to improve the estimate. PA2 is usually the method of choice.

STEP II.6c. CLASSICAL FACTOR ANALYSIS

A printout for these two methods from FACTOR in SPSS is given in Table II.6c. The data base used was the air pollution data for 456 cases at four stations. Results from the two methods are combined for comparison. Initially, the program lists simple correlation coefficients like those obtained with CORREL in SPSS (not shown here). All correlation values for a variable with itself should equal one. For PA1 analyses, these are usually left as one in all cases and are used as the estimate of communality. In PA2, these are replaced by a value that represents the communality present. The squared multiple correlation between that variable and the rest of the variables is used (recall this statistic was used in regression analyses in Step II.3f). These are shown as the initial communality estimates in the table. Note that results for PA1 are given in the left side of the table and PA2 on the right. All communalities for PA1 are estimated as 1.00.

For PA2, the estimated communality for MN = 0.754. This is the initial estimate of the amount of MN variance due to the "common" part. Since the variable was standardized before usage to have a total variance of 1.0, the remaining 0.246 is estimated to be due to the unique part.

TABLE II.6c
Factor Analysis from FACTOR (SPSS)

	PA1			PA2

Initial Communality Estimate for the Variables

VARIABLE	ESTIM COMMUNALITY	FACTOR		VARIABLE	ESTIM COMMUNALITY
MN	1.00000	1		MN	.75437
CU	1.00000	2		CU	.20311
FE	1.00000	3		FE	.68601
PB	1.00000	4		PB	.73123
PART	1.00000	5		PART	.80191
ORG	1.00000	6		ORG.	.70914
SUL	1.00000	7		SUL	.17778
NIT	1.00000	8		NIT	.20455
CL	1.00000	9		CL	.38212
CO	1.00000	10		CO	.59377

Initial Factors Extracted

EIGENVALUE	PCT OF VAR	CUM PCT
3.89657	39.0	39.0
1.89254	18.9	57.9
1.24839	12.5	70.4
.82176	8.2	78.6
.71563	7.2	85.7
.58665	5.9	91.6
.27718	2.8	94.4
.26363	2.6	97.0
.17772	1.8	98.8
.11993	1.2	100.0

Retained Factors

	FACTOR1	FACTOR2	FACTOR3			FACTOR1	FACTOR2	FACTOR3
MN	.67402	-.61737	.14396		MN	.67888	-.58008	.09281
CU	.27723	.48385	.52162		CU	.22577	.31347	.42735
FE	.64795	-.64245	-.08785		FE	.64974	-.56773	-.17830
PB	.79637	.39268	-.21354		PB	.78141	.43904	-.10178
PART	.79622	-.48640	.09330		PART	.81125	-.46104	.07555
ORG	.74943	.41947	-.24973		ORG	.72138	.44241	-.12443
SUL	.48034	-.03658	.16901		SUL	.39127	-.01037	.09530
NIT	.21500	.22378	.75882		NIT	.17436	.11950	.51172
CL	.63905	.20644	.21011		CL	.55358	.15923	.22836
CO	.65606	.45138	-.42740		CO	.62702	.47654	-.30072

Final Communality

	COMMUNALITY				COMMUNALITY
MN	.85618			MN	.80600
CU	.58305			CU	.33186
FE	.84030			FE	.77627
PB	.83400			PB	.81371
PART	.87926			PART	.87639
ORG	.79996			ORG	.73159
SUL	.26065			SUL	.16228
NIT	.67212			NIT	.30653
CL	.49515			CL	.38395
CO	.81683			CO	.71068

FACTOR	EIGEN VALUE	PCT OF VAR	CUM PCT
1	3.60622	61.1	61.1
2	1.62493	27.5	88.7
3	.66810	11.3	100.0

Varimax Rotation

	FACTOR1	FACTOR2	FACTOR3			FACTOR1	FACTOR2	FACTOR3
MN	.92163	.05498	.06133		MN	.89114	.06226	.08937
CU	-.10158	.22973	.72108		CU	-.05522	.19431	.53949
FE	.89359	.12713	-.16015		FE	.85213	.15020	-.16607
PB	.22287	.86922	.16967		PB	.21777	.84768	.21847
PART	.90203	.23622	.09905		PART	.89572	.23408	.13885
ORG	.16726	.86804	.13595		ORG	.17309	.81723	.18375
SUL	.37103	.23618	.25923		SUL	.28088	.22509	.18088
NIT	.07569	-.06656	.81361		NIT	.05267	.00561	.55112
CL	.30548	.46464	.43121		CL	.27510	.39380	.39139
CO	.05909	.90125	-.03287		CO	.07684	.83948	.00659

The initial factors are then extracted, and are listed under "INITIAL FACTORS" in the table. For PA1 (like KARLOV in Step II.6b) ten factors are found, each being a linear combination of the variables in the set. Each factor has an associated eigenvalue that is related to the amount of variance information retained in that factor. Note that these values for PA1 are similar to those from KARLOV. The FACTOR program defaults at this point to retain only those factors having an eigenvalue greater than 1.0. Since each individual original variable began with this much variance, the only eigenvectors retained are those containing at least as much variance as one of the original variables. This can be changed by the user. The eigenvectors are listed below this for the three retained factors for PA1. As with KARLOV, the eigenvalue 3.89657 for factor one is obtained by

$$(.674)^2 + (.277)^2 + (.648)^2 + \cdots$$

Under this listing on the table, the communality retained for each variable in these three factors is listed. Considering only the first three factors, MN retains

$$(.674)^2 + (-.617)^2 + (.144)^2 = .856$$

of its variance. It has, therefore, lost .144 in the seven unretained factors. Note that the three retained factors describe most of the variance for MN, FE, PB, PART, ORG, NIT, and CO, but do not do as good a job for CU, SUL, or CL. As with KARLOV, a plot could be made of the first two axes which would retain a total of 57.9% of the variance.

PA2 (on the right side of the table) determines the number of factors to be extracted from the data in an identical manner to PA1. Therefore, three factors will again be extracted. The main diagonal elements of the correlation matrix are replaced by the estimated communality values listed and factors are extracted. The variance of each variable accounted for by these three factors becomes the new communality estimate. The diagonal elements of the matrix are replaced by these new values and the process continues until the differences between two successive steps are negligible. If PA1 were used and estimates of the communality specified to be other than 1.0 for all values, this process would be done without the iteration step.

In the table, PA2 lists the final factors extracted and the final adjusted communality for each variable. For MN in PA2, the "common" was originally estimated to be 0.75437. After the three factors were extracted, the MN communality was

$$(.679)^2 + (-.580)^2 + (.093)^2 = .806$$

The remaining MN variance of .194 was not described in these three factors. Note that the variances for MN, FE, PB, PART, ORG, and CO were well retained in the common. In other words, these variables were well defined by the three extracted factors. The variables CU, SUL, NIT, and CL were not.

The total variance (assuming that the variance for each of the ten variables was originally 1.0) for the problem was 10.0. The communality retained in the three factors was

$$0.806 + .332 + .776 + \cdots = 5.899$$

The remaining 4.101 was assigned to the unique part. And of the 5.899 explained

$$(.679)^2 + (.226)^2 + (.650)^2 + (.781)^2 + \cdots = 3.61$$

3.61 was retained in the first factor. This calculates to be 3.61/5.899 or 61.1% of the communality. The second factor, similarly, explained 27.5%, and the third 11.3%.

PA2, therefore, indicates that MN (V_1) spreads out over the factors as

$$V_1 = .679F_1 + (-.580)F_2 + .093F_3 + d_1U_1$$

This indicates that MN is most important in the first two factors. The amount of variance present for MN that is accounted for in factor 1 is given by $(.679)^2 = .46$, or 46% or its total variance is in factor 1. Likewise factor 2 accounts for $(-.580)^2$ or 34% and factor 3 is $(.093)^2$ or 1%. These "common" factors account for $46 + 34 + 1 = 81\%$ of the variance. The "unique" portion contains the other 19%. Since the variance in the unique portion is given by 0.19, the value of d_1 for MN would be $\sqrt{.19} = .44$. Therefore, the final equation for MN would be

$$V_1 = .679F_1 - .580F_2 + .093F_3 + .44U_1$$

Next, one should check to see if this factor analysis solution retained the correlation between the original variables. Remember that the assumption was made that all of the correlation was due to the common and not the unique. Therefore, if one is to consider MN and FE, whose overall simple bivariate correlation is 0.757 (from CORREL), it must be retained in their correlations in the three factors. Considering MN variable 1 and FE variable 3, this means that $(r_{13})_t$ or the total correlation is given by the sum of r_{13} due to each of the three factors $F1$, $F2$, and $F3$ or

$$(r_{13})_t = (r_{13})_{F1} + (r_{13})_{F2} + (r_{13})_{F3}$$

and r_{13} is given by multiplying the factor loadings for the two variables in the given factor. Therefore

$$(r_{13})_t = (.679)(.650) + (-.580)(-.568) + (.093)(-.178)$$
$$= .44 + .33 - .02 = .75$$

The correlation has indeed been preserved in the common parts of the factor analysis.

These are the common mathematical calculations necessary to interpret factor analysis results. We can find the eigenvectors associated with those transformations accounting for a maximum amount of the variance in a minimum amount of axes or factors. For the PA1 result, these two factors could then be plotted to represent the best possible two-dimensional plot of linear combinations of the variables to account for the variance. With PA2, the background or unique portion of the variable has been removed first, and then the remaining variance has been represented in a minimum number of factors. It has been assumed that this correlation between the variables does reflect some underlying factors that have led to their values.

One more step may be used in the factor analysis problem. As with regression analysis, the solution obtained in a factor analysis is not unique. There are other statistically equivalent ways to represent this data and there is no way to determine which representation is "best." This is where the scientist must add input to the problem. Not all mathematically correct results lead to theoretically useful results. The representation chosen is then the one that best summarizes the data in a theoretically meaningful way.

FACTOR ROTATIONS

The method used to study the other equivalent representations of the factors derived is called a rotation. Transformations of this type can be made to produce equivalent factors. Two major options will be discussed for this step. Orthogonal rotations will be found to be mathematically simpler while oblique ones may be more realistic. The difference lies in the fact that orthogonal factors will be uncorrelated while oblique ones may be correlated. Three common orthogonal rotations are varimax, quartimax, and equimax.

Table II.6c gives results of the varimax orthogonal factor rotation available in FACTOR of SPSS. Varimax rotations concentrate on simplifying the columns of the factor. This makes the complexity of the factors a minimum. One would then expect each factor to have heavy loading in only a very few variables. Stated another way, it attempts to represent each factor with only a few heavily loaded variables.

In quartimax rotations, the goal is to make the complexity of the variables a minimum, which means that each variable loads high in only a few factors. Therefore, for example, MN is described as belonging to only a few factors.

In equimax rotations, a compromise between the quartimax and varimax rotation is sought. An attempt is made to simplify both the factors and the variables.

In Table II.6c, Part D, results from the varimax rotations are shown. The other two rotations were tried and similar results were obtained. Factor 1 weighs heavily in MN, FE, and PART; factor 2 in PB, ORG, and CO; and factor 3 in CU, NIT, and CL. Factor 3 is not as well defined (weightings for the variables are not as high) as for the other two factors. This can be expected since the eigenvalue for this factor is only .688 or 11.3% of the total. Note how different these results are from the original unrotated factors. In the unrotated case, this relationship between variables is difficult to obtain, since variable loadings tend to be spread out among all factors. It is our suggestion to experiment with the various options for rotation, but do try them. Keep in mind, however, that the computer cannot inject scientific explanations into the obtained results. It can only be used to obtain the various rotations. The user must then decide which is best for describing his individual problem.

Oblique rotations will not be discussed at this point. Details may be found in the SPSS manual or various factor analysis textbooks.

The results obtained for this problem suggest two major and one minor pollution sources. The first source (factor 1) is heavily loaded by the variables MN, FE, and PART. This is very likely to be due to the desert dust around Phoenix. The second factor, loaded in PB, ORG, and CO, could come mainly from automobile exhaust in the area. The third, less well defined factor, could be related to the mining operations in the area.

It would be instructive, at this point, to look also at the fourth, and maybe even the fifth, factors as well. It would also be very useful to study factor analysis results for each sampling station individually, since each source should have different importance at each.

In Step II.6, various methods for reducing the data base have been shown to include only data (which may represent variables, linear combinations of variables, or some type of intervariable relationships) that best represent the problem to be considered. This can then be utilized to determine which variables are most important to analyze when other data are collected. This can reduce the number of analyses necessary, therefore resulting in a savings of time and possibly money for the scientist.

UNDERLYING VARIABLE FACTOR ANALYSIS

Infometrix, Inc., the source for the ARTHUR program, also supports an ARTHUR-compatible program, UVFA, that goes beyond the principal components analysis available in ARTHUR. UVFA allows the analysis of data by both linear and nonlinear factor analysis methods. The intrinsic dimensionality of the data can be studied using the program, which also features a variety of other pattern recognition techniques such as multidimensional scaling, parametric mapping, varimax and oblique rotations, eigenvector interpretations, back-transformations and error calculations, hierarchical factor clustering, nonlinear mappings, and other various plotting routines.

II.7. DATA MANIPULATIONS

INTRODUCTION

In this step, we will briefly consider some of the data modifications that can be made now. We are not concerned with statistical and mathematical manipulations (see Appendix IX for those types of modifications). We will be more concerned with ideas on use of the computer for sorting data, rearranging cases, choosing variables, and performing these types of modifications. Many are built into the computer packages and much time can be saved with minimal amounts of effort.

To this point, the examples used in this work were presented to the computer in the simplest way possible. Chapter III is concerned only with demonstrating these data introduction methods. However, after one becomes familiar with the typical runs, adding the data manipulation capabilities considered here is not difficult. The goal is to be able to change or choose certain data in the overall data base without the need for retyping or physically sorting out the cards. Only an introduction to these will be made. The respective program manuals give implementation details.

SPSS CAPABILITIES

In SPSS, a wide variety of techniques is available. These are usually implemented with a control card in the introductory data definitions. Brief consideration of each will be made here with further details available in the SPSS manual.

"RECODE" allows the user to manipulate the variables in the file. Continuous variables can be grouped into discrete categories for further analyses. Variables can be converted from alphanumeric variables to true numeric ones so that mathematical manipulations may be performed. A choice can be made as to whether the change is temporary (only to be used for that particular analysis) or permanent. If the latter is true, the change is used for all analyses of the data in that run and can be retained on "SAVE FILE" for further analyses.

The "COMPUTE" facility allows the user to perform arithmetic calculations on the variables in the file, and store these for the data input for other subroutines. Simple FORTRAN expressions are used to designate the desired transformation. The newly created variable is added to the variable list as the last one in the file. Addition, subtraction, multiplication, division, square roots, logarithms, geometric functions, and truncations are possible. These may be performed on either a single original variable or a combination of original variables. This modification, like "RECODE," can be used as a temporary or permanent change in the file structure.

"IF" cards can be used to change variables subject to some condition being true. The new variable will be calculated only if the condition immediately after the "IF" is true. The new variable can be calculated in a manner similar to the "COMPUTE" expression. If the condition is false, variables keep their previous values. The condition to be met can contain relationships such as "greater than," "less than," "equal to," "greater than or equal to," etc. Also, logical operators that connect two relationships such as "and," "or," or "not" can be used. Therefore, a large variety of combinations exist. An example of such a combination would be to add variables A, B, and C to form variable D "IF" (a) variable A is greater than or equal to B (b) and B is less than C (c) or C is equal to zero. The correct implementation of this can be found in the SPSS manual, Chapter 8. The "IF" card can be permanent or temporary as well.

A "COUNT" card can be used to determine the number of occurrences of a particular characteristic. New variables can be created from these. Several variables may be counted simultaneously. Oftentimes, more storage space in the computer may be necessary to handle all of the new variables. An "ALLOCATION" card is used for this purpose. Data may also be selected by use of the computer. A "SAMPLE" card can be used to obtain a random sample from the file. The percentage of the file to be included is specified. This can be done temporarily or permanently. A second method uses a "SELECT IF" card, which then lists a logical expression to be considered. If the expression holds for a given case, the case is selected to include in a subset of the data. If more than one "SELECT IF" card is used, cases will be chosen if any of the logical expressions hold true.

Many methods for creating and changing file systems in SPSS are available. The common method (shown in Chapter III) is to include a "SAVE FILE" card in the computer run. This data set will be stored for the user under the file name given. It can be retrieved with a "GET FILE" card. If a file is saved to be used at a later date, it also can be modified. Variables can be added, deleted, kept, or recorded. Data lists can be added to the present set or files of separate data may be merged. Cases may be added or sorted. Subfiles may be deleted. All of these manipulations are discussed in Chapter 11 of the SPSS manual. The rationale for including such a large variety of possibilities is to aid the user in manipulations for large data bases. If the number of cases and/or variables in the base is large, handling such rearrangements by hand becomes an impossible task. SPSS has therefore made many options for such manipulations available through reasonably easy access.

BMDP AND ARTHUR ALTERATIONS

Transformations in BMDP are usually done in one of two places. The first is a "TRANSFORM" paragraph in a given BMDP program. In this method, the user can combine, scale, or recode variables. New variables can be created. Cases can be selected or deleted according to certain criteria. Random subsamples can be drawn from the data. These methods are all very similar to those in SPSS. Details can be found in the BMDP manual. BMDP1S is a program that is dedicated to multipass transformations. At each pass, the data can be altered or edited. Many options are available to the user for this. Changes can then be saved in a file (see Chapter III).

In ARTHUR, modifications are available through subroutine CHANGE, which allows feature, category, pattern, and file changes to occur. Variables may be combined, deleted, or added. Category definitions may be altered and patterns changed to allow new definitions for training and test set inclusions. An unlimited number of changes are available. Changes may be temporary for that run or filed as a permanent change for the data base. Subroutine TUNE will combine variables using mathematical manipulations found to be commonly useful. Through TUNE, new variables are added to the data set that are calculated by

a. the inverse of each original variable in the set;
b. combinations of two variables multiplied by each other;

    c. all possible ratios of variable pairs;

    d. the inverse of (b) given above;

    e. the square of each variable.

Even for a relatively small number of original variables, the number created in this subroutine is enormous. Only a relatively small number of the new variables are probably useful, and can be found by utilizing Step II.6a (SELECT) immediately following a TUNE. (A prerequisite for SELECT is SCALE to create standard scores for the variables—see Appendix IX. Therefore, you must do TUNE, SCALE, and then SELECT.)

It is not the intention of this section to consider data modifications in detail. With the above information on what might be considered, the manuals will give details on the implementation.

III

IMPLEMENTATION

INTRODUCTION

In this chapter, the "how to do it" information will be given. Enough detail will be provided to enable the beginner to obtain results. This chapter is to be used in conjunction with details in the SPSS, BMDP, and ARTHUR manuals. It will be assumed that the user knows how to punch computer cards, or code information into the CRT, or other type of computer input device. The user must have available the information on how to access the computer to be used for the analysis. This information can be obtained at the user's computing facility. The user must find out how to assign SPSS, BMDP, and ARTHUR at his given facility. This chapter will be concerned with the cards or commands necessary to complete successfully runs for the three programs after they have been assigned. Throughout this chapter, we will discuss only computer cards as the input medium. Other methods will utilize similar package commands. See the manual for details.

One note to be made is that with SPSS and ARTHUR, the entire package is assigned by a given card, and all subroutines within that package can then be accessed in a single computer run. With BMDP, each method is a program and not a subroutine within a main program. Therefore, each must be accessed individually. Separate computer runs are needed for each type of analysis. The rest of the discussion will focus on necessary steps after the programs are successfully accessed.

With all three programs, a sequence of control cards is needed. These dictate various instructions to the computer for the analysis. Each program requires a certain format that must be used for preparing the control cards. These will be discussed for each package individually in the next three sections.

It will be assumed that the data to be analyzed have already been placed on cards in the correct format. Instructions for this are given in Appendix III. The control cards will then be placed in the data deck after the cards necessary for accessing a particular program and before the data cards. Also, it is necessary to place cards after the data deck to indicate what (other) analyses are to be performed or to indicate that the instructions are finished.

The entire data deck is then submitted to the computer, using the method appropriate for the given computing facility. The answers will be returned on a printout containing results similar to those in the tables in Chapter II.

There are other methods for accessing the computer. Card decks are not necessary. Cathode ray tubes (CRTs) can be used. Data decks may be made into files which are added to the runstream at the appropriate position. Our goal is to familiarize the user with the programs only, and not with the variety of implementations possible. A minimal number of modifications will allow access by these other methods as well. Instructions can be obtained from the computing facility. These modifications are quite dependent on the computer that is being used. Therefore, details for these will not be addressed.

Section III.1 describes typical control card decks for SPSS analyses. These are prepared identically for all of the different programs. A discussion of the individual programs present is given in III.4, which will concentrate on the most common methods and options that are used for each. Many others do exist and are explained in the manual.

A discussion of the control cards for BMDP programs, and the options available, is given in Section III.3. For this, the control cards (although similar) are not identical for each method, so the discussion will center on typical ones that are usually needed. Methods for accessing the individual programs are discussed in III.6.

Sections III.2 and III.5 are devoted to descriptions of control cards and program cards for ARTHUR. Options for the subroutines will be studied.

If the data format specifications described in Appendix III are used, the same data deck may be used for all three programs. The only changes necessary are in the control cards and program cards. Methods from all three packages can therefore be used, with little extra effort, to obtain a complete statistical study for the data.

For the examples in this chapter, the air pollution data will be used. The tables for each will show both a listing of the cards read into the computer, and the first few pages of the computer printout obtained from the run.

III.1. TYPICAL SPSS RUNS

INTRODUCTION

Various SPSS analyses will be discussed. The first, given in Table III.1a, is the simplest. This run consists of the 456 patterns of data being considered as a single group. No category structure is used. Ten variables (MN, CU, FE, PB, PART, ORG, SULF, NIT, CL, and CO) are measured on each. This run could be used for unsupervised analyses. In this case, the example used demonstrates input into PEARSON CORR for the data considered as a single set.

The second example, given in Table III.1b, uses the same data base, but divides the data into four categories (MC, MS, NS, and SC). A PEARSON CORR is again run, with the goal this time being to compare the intervariable correlations considering the categories individually, and then as a whole.

When typing SPSS cards, each is divided into two sections. Columns 1–15 (called the control field) are reserved for control words and columns 16–80 are for the specification field. The control words indicate what procedure is to be performed, and the specification indicates details for the analysis (and on what variables). This will be explained further. Remember always to utilize this format. The control word must begin in column 1 and will never exceed 15 characters. Spelling and spacings are critical. The specification field does not have to adhere to a rigid format. If more space is needed for it, the specifications can continue on a second (or even third) card, as long as the second and subsequent cards leave columns 1–15 blank.

Keywords may be used to designate variables such as "TO," "BY," "WITH," and "THRU." These are discussed in the examples (and in the SPSS manual). In SPSS, blanks and commas can be used interchangeably and may be inserted between specifications if desired. These cannot be used in the control field. Delimiters such as parentheses, slashes, and equal signs can be used only when a given control card specifies their usage.

CARD DECK

Table III.1a shows a card list and a partial printout for an SPSS computer run. The pollution data were used. All 456 cases were considered as a single group. Note that the printout lists the cards as it encounters them. This aids in determining errors in the control card deck.

The first card shown is optional. In columns 1–15 are the words "RUN NAME" with a blank between the two. Beginning in column 16 is a name for this particular computer run. Up to 64 characters in any format can be used. Although this card is optional, it is suggested that it be included. This name will then be printed at the top of each page of the printout.

The second card is a "VARIABLE LIST" control card. Variable names in SPSS can consist of up to eight characters with the first being alphabetic. Each name (which must be unique from the others) is then used throughout the analysis to refer to those data. The list must be in the order that the variables will be encountered on the data cards. In this example, ten variables were listed. The names given (MN, CU, FE, PB, PART, ORG, SULF, NIT, CL, CO) were abbreviations for the chemical species analyzed. Meaningful abbreviations allow the user to remember the origin for each variable. The data base will then have to consist of values for at least these ten variables in correct order (other variables may also be present if the input format ignores them— see card #5). The variable names can be separated by commas or spaces.

The third card is "INPUT MEDIUM." For our example, the specification for this is "CARD" since the data are given on computer cards. This control card signifies where to find the data. Other possibilities such as tapes or disks will not be considered here (see SPSS manual).

TABLE III.1a

SPSS Run with No Subfile Structure

```
1234567890123456789012345678901234567890123456789012345678901234567890

RUN NAME          PEARSON CORR/  NO SUBFILES/  CHAPTER III Ex. III-1a

VARIABLE LIST     MN,CU,FE,PB,PART,ORG,SULF,NIT,CL,CO

INPUT MEDIUM      CARD

N OF CASES        456

INPUT FORMAT      FIXED(9X,4X,2F5.0,2X,6F7.0,F5.0,F6.0)

MISSING VALUES    MN,CU,CL(99999)/FE TO NIT(9999999)/CO(999999)

PEARSON CORR      MN TO PART

OPTIONS           3

STATISTICS        1

READ INPUT DATA

        data cards are placed here.

FINISH
```

Part B First Page of Output

SPSS 6.03   NOVEMBER 4, 1976

SPACE ALLOCATION FOR THIS RUN..

TOTAL AMOUNT REQUESTED   9000 WORDS

DEFAULT TRANSPACE ALLOCATION   1125 WORDS

MAX NO OF TRANSFORMATIONS PERMITTED   37
MAX NO OF RECODE VALUES   150
MAX NO OF ARITHM.OR LOG.OPERATIONS   300

RESULTING WORKSPACE ALLOCATION

1.   RUN NAME   PEARSON CORR/ NO SUBFILES/ CHAPTER III Ex. III-a1
2.   VARIABLR LIST   MN,CU,FE,PB,PART,ORG,SULF,NIT,CL,CO
3.   INPUT MEDIUM   CARD
4.   N OF CASES   456
5.   INPUT FORMAT   FIXED(9X,4X,2F5.0,2X,6F7.0,F5.0,F6.0)

ACCORDING TO YOUR INPUT FORMAT, VARIABLES ARE TO BE READ AS FOLLOWS

VARIABLE	FORMAT	RECORD	COLUMNS
MN	F 5. 0	1	14- 18
CU	F 5. 0	1	19- 23
FE	F 7. 0	1	26- 32
PB	F 7. 0	1	33- 39
PART	F 7. 0	1	40- 46
ORG	F 7. 0	1	47- 53
SULF	F 7. 0	1	54- 60
NIT	F 7. 0	1	61- 67
CL	F 5. 0	1	68- 72
CO	F 6. 0	1	73- 78

THE INPUT FORMAT PROVIDES FOR   10 VARIABLES.   10 WILL BE READ
IT PROVIDES FOR 1 RECORDS ('CARDS') PER CASE.   A MAXIMUM OF   78 'COLUMNS' ARE USED ON A RECORD.

6.   MISSING VALUES MN,CU,CL(99999)/FE TO NIT(9999999)/CO(999999)
7.   PEARSON CORR MN TO PART
8.   OPTIONS   3
9.   STATISTICS   1
10.   READ INPUT DATA

TABLE III.1b
SPSS Run with Subfile Structure

```
Part A   Data Listing (pre-Release 8 run)
123456789012345678901234567890123456789012345678901234567890123 4567890

RUN NAME           PEARSON CORR/  SUBFILES  /  CHAPTER III Ex. III-1b
VARIABLE LIST      MN,CU,FE,PB,PART,ORG,SUL,NIT,CL,CO
INPUT MEDIUM       CARD
SUBFILE LIST       MAR(114)  MESA(114)  NSCT(114)  SCOT(114)
INPUT FORMAT       FIXED(9X,4X,2F5.0,2X,6F7.0,F5.0,F6.0)
MISSING VALUES     MN,CU,CL(99999)/FE TO NIT(9999999)/CO(999999)
RUN SUBFILES       EACH
PEARSON CORR       MN,CU,FE,PB,PART
OPTIONS            3
STATISTICS         1
READ INPUT DATA

    data cards are placed here
RUN SUBFILES       ALL
PEARSON CORR       MN TO PART
OPTIONS            3
FINISH


Part B   Data Listing with Release 8 Options

123456789012345678901234567890123456789012345678901234567890123 4567890

RUN NAME           PEARSON CORR/  SUBFILES/  CH.III Ex. III-1B/  RELEASE 8 OPTIONS
VARIABLE LIST      GROUP,MN,CU,FE,PB,PART,ORG,SUL,NIT,CL,CO
INPUT FORMAT       FIXED(9X,F4.0,2F5.0,2X,6F7.0,F5.0,F6.0)
MISSING VALUES     MN,CU,CL(99999)/FE TO NIT(9999999)/CO(999999)
SORT CASES         GROUP/SUBFILES
RUN SUBFILES       EACH
PEARSON CORR       MN,CU,FE,PB,PART
OPTIONS            3
STATISTICS         1
READ INPUT DATA

    data cards are placed here

END INPUT DATA
RUN SUBFILES       ALL
PEARSON CORR       MN TO PART
OPTIONS            3
FINISH
```

The fourth card is the "N OF CASES" control card. This informs the system of the number of patterns of data to expect. In this example, 456 patterns were included.

With Release 8 of SPSS, the "INPUT MEDIUM" specification "CARD" is assumed by default and need not be included. Also, for card input, a "N OF CASES" card can be deleted if an "END INPUT DATA" card is inserted after the last data card in the data set. This is shown and discussed later in Table III.1b, Part B.

To this point in the control card deck, the computer has been informed to expect card input containing 456 cases, each having data for ten variables. The system must then be told the format necessary for reading these cards. This is done on card #5, the "INPUT FORMAT" control card.

DATA FORMAT STATEMENT

The format for the air pollution data and instructions on methods for data format are given in Appendix III. In SPSS, case names are not recognized by the system. These are only of use for card identification purposes. SPSS assigns sequence numbers to each case and refers to these as the variable "SEQNUM." Therefore, the case name given in ARTHUR-compatible format is deleted from the format for SPSS. The ID number in the format is also deleted. SPSS has a different method for category designation. This will be shown in the second example. Therefore, the category format can also be ignored. The previous ARTHUR format (I1, 2A4, F4.0, 2F5.0, 2X, 6F7.0, F5.0, F6.0), therefore, becomes (9X,4X,2F5.0,2X,6F7.0,F5.0,F6.0) in SPSS. The "9X,4X" could have been given as "13X".

SPSS can handle formats that are either in fixed columns or free-field. In free-field formats, the same columns on successive data cards need not contain values for the same variable. Only the order of variable specification must be identical. For instance, in the first case, variable #1 may take columns one through three while in the second case it could occupy columns 1–5. This is not ARTHUR-compatible. We will, therefore, concentrate on fixed column formats. The key word "FIXED" must be the first word in the specification field of the "INPUT FORMAT" card. This is followed by the format for the data, listed in parentheses. A comma is required between each format specification as is seen in the example. Blank spaces may be used as needed.

The variable list and format specification are paired from left to right. The printout lists this pairing so the user can check the two. For this example, MN is given a format of F5.0 which is found in columns 14–18 of the first record (or card). Similar results are shown for each variable. If the two do not match, SPSS gives an error statement at this point. At the end of the output, error numbers and descriptions are listed to aid the user in identification of the corrections necessary.

MISSING DATA

The sixth card is necessary if any of the variables have any data missing. If not, this card is omitted. If missing data are present, the user must decide how this is to be coded. Appendix VIII addresses this problem. The sixth card in this example, the "MISSING VALUES" card, indicates which variables have missing data, and how these are coded for each. Since ARTHUR-compatible format is suggested for use, the missing value designation is made by filling each column for that variable with 9's for the cases where the data are missing. In the example in Table III.1a, data were missing, so the "MISSING VALUES" card was included. The format specification included five columns for MN, CU, and CL; six columns for CO; and seven columns for FE, PB, PART, ORG, SULF, and NIT. On this specification card, the value (or values) coded to represent missing data are placed in parentheses following the variable. Up to three values can be specified for each variable. In our case, only the 9 specification is used. Therefore, card #6 shows, in the specification field, the values 99999 for MN, CU, CL; 999999 for CO; and 9999999 for FE to NIT. The other variables did not contain missing data. Note the use of the keyword "TO." This means that this specification is for all variables from FE to NIT. Note also that each variable or variable list with certain missing value name(s) is separated from the next one by a slash mark. This can be continued onto a second card (not using columns 1–15) if necessary.

These six control cards are the only ones needed for the type of data in the example. Many other optional ones are available and aid the user in identifications on the output. Labels, documentation, and extended descriptions can be added. The SPSS manual describes these.

STATISTICAL PROCEDURES

In SPSS, the first statistical procedure to be used is listed next. In Table III.1a, cards 7–9 are used for this. The tenth card, "READ INPUT DATA," is then used to indicate that the control specifications are finished, and the data cards are following immediately. Therefore, the data begins as card #11 and continues until all data are added to the problem. The computer knows that 456 cases of data are present. This was indicated on the "N OF CASES" card. After the last data case, more statistical procedure cards can be added as desired although none are necessary. When all of the statistical procedures that are desired have been added, a "FINISH" control card is added as the last SPSS card in the deck. A normal exit from the computer must then be indicated, using the cards necessary for a given computer facility.

A description of the statistics cards (numbers 7–9 and those after the data deck) will be given next. These are often referred to as the task-definition cards. Instructions for specific calculations desired are described on these cards. The selection of the subprogram in SPSS to be used is indicated on the procedure card, the first in a set of three task-definition cards. Each program has a corresponding procedure card. The control words for these are the program names given in Appendix IV and designated throughout this book as small capital letters. To access a desired routine, the program name is listed on the procedure card in columns 1–15. The specification field is then used to define what variables, parameters, and arguments are to be considered for the problem. These are specific to the subprogram, and are described in Section III.4.

After the procedure card, an "OPTIONS" and/or "STATISTICS" card can be used. The "OPTIONS" card provides additional information for the calculations desired. A choice of various ways of handling the data is given. The choices are indicated by option numbers given in the specification field of the card. The "STATISTICS" card indicates choices among available statistics to be included in the analyses. The word "STATISTICS" is used in the control field and the choices are listed by number in the specification field. Both "OPTIONS" and "STATISTICS" card are defined by the various procedures and will be described in Section III.4 (or in the SPSS manual).

Therefore, in this example, a PEARSON CORR is to be performed on variables MN to PART utilizing option #3 and statistic #1. The "READ INPUT DATA" card follows, and then the data cards. A "FINISH" card indicates that this is the only procedure desired. The computer is then exited by the appropriate means. Note the printout reproduces all of the information from the control cards, to aid in error identification.

CATEGORY DEFINITIONS

Example III.1b utilizes the same data cards as III.1a. In this case, however, the four sampling stations are to be considered as separate categories, and correlation coefficients are desired for each individually and then the data base as a whole. In the 1976 version of SPSS, categories are not given number designations as in ARTHUR. In this case, "subfiles" are created for each. It is important that the data be sequenced such that all patterns belonging to the same category are grouped together. This means that, in the example, all cases for MC are listed first followed by those for MS, then NS, and finally SC.

With Release 8 of SPSS, it is also possible to define subfiles on a "SORT CASES" control card, placed in the deck before specifying the first procedure. Therefore, it is no longer necessary to sequence the data if this option is chosen.

Table III.1b shows the data list and partial printout for this run (Part A), and the option available in Release 8 (Part B). In Part A, the first three cards are identical to those from the first example. The change occurs in the "N OF CASES" control card. For data with subfile structures, this card is omitted. In its place, a "SUBFILE LIST" is used to define the data division. In the specification field of this card, the system is informed how to divide the data into subfiles. The name of each subfile is listed followed by the number of cases in each in parentheses. Only the first four letters of the name are recorded by SPSS to define a variable named "SUBFILE." Therefore, those four must be unique for each category. In this case, four categories (or subfiles) are defined: MAR, MESA, NSCT, and SCOT. Each contains 114 cases. These are found in the data deck with the 114 for MAR listed first, the 114 for MESA second, and so on. This is equivalent to 1.0, 2.0, 3.0, and 4.0 listed in the category field of the ARTHUR format. Note that these category numbers can be used and recognized in Release 8 of SPSS.

In all examples, a card is inserted to describe missing values. In Table III.1b with the subfile list, the computer must then be told how to consider the subfile structure. A "RUN SUBFILES" card is used for this. The subfiles can be considered individually. The specification field then contains the word "EACH." For the run in Table III.1b, this was the case. The three task-definition cards follow, using a format similar to that described in III.1a. Therefore, for this example, a Pearson product–moment correlation matrix is calculated for each station (or subfile) individually. These four can then be compared to see how correlations vary between stations. The card after the task definition ones is the "READ INPUT DATA" card followed by the data deck, as was the previous case.

Now assume that the correlation matrix is desired for the data base as a whole as well. After the data cards, other statistics can be specified. In this case, this additional procedure is desired. First, the computer must be told that all of the subfiles are now to be considered as a single group. For this, a "RUN SUBFILES" card is used with the word "ALL" in the specification field. The second set of task definition cards is then included. In this example in Table III.1b, a correlation table is calculated on variables MN to PART considering the entire data base at once using option 3. No additional statistics are listed (indicated by no "STATISTICS" card). This procedure is then completed. After the second set of task-definition cards, a "FINISH" card is used to indicate the end of the procedures desired. Therefore, these two are completed, and the run is finished. After each procedure on the printout an error diagnosis would be given.

A third option exists for the "RUN SUBFILES" card. A particular subset of subfiles may be desired for a certain process. They can either be considered individually or as an entire subset. These options can be chosen in the specification field of the "RUN SUBFILES" card. For instance, if the user desired to calculate a correlation matrix for MC and MS individually and then for NS and SC combined, the specification field would be "(MC), (MS), (NS,SC)." Those subfiles enclosed together in parentheses are combined into a single unit. Those listed individually in parentheses are considered individually. These two options can be combined in a single "RUN SUBFILES" card as seen here.

The specifications on this card are then in effect until another "RUN SUBFILES" card is encountered. Changes in this are placed in the data deck immediately before the procedure card for the programs in which the changes are to be made.

In Part B of Table III.1b, the optional changes available for Release 8 of SPSS are shown. The "VARIABLE LIST" now includes an additional variable, GROUP, which designates classification. In the format statement, this variable is added as an F4.0 specification in columns 10–13. In Appendix III, it can be seen that these columns contain a 1.0 for MC, 2.0 for MS, 3.0 for NS, and 4.0 for SC. The "SUBFILE LIST" card is not needed, but is replaced by a "SORT CASES" card which specifies that the variable GROUP is to be used to designate the subfiles.

Note that the "INPUT MEDIUM" card has been removed since CARD is assumed by default. No designation as to the number of cases has been made. Instead, an "END INPUT DATA" card is placed after the last data piece to signify the end of the data set.

FILE SAVING

SPSS can also save a data file created in a run. A location for the saved file must be assigned. The method for this is dependent upon the computer used. Examples are given in the appendices of the SPSS manual. The control and task definition cards for such a run are similar to those in Table II.1b.

One change necessary is to name the file under consideration. Since the data file is to be accessed again later, it must be given a name. This occurs in the "FILE NAME" card placed between the "RUN NAME" and "VARIABLE LIST" control cards. The card lists a file name in the specification field of the card. It must be less than or equal to eight letters long, the first of which is alphabetic.

After the last task definition card and before the "FINISH" card, a "SAVE FILE" card is inserted. No additional specifications are necessary (although it can be used to rename or relabel the file—see manual). The printout obtained is identical to that in Table III.1b with the addition of the results of the save file procedure.

The file can then be retrieved for future analyses. Again, the assigning of the file is computer dependent and will not be detailed here. SPSS only needs a "GET FILE" card to retrieve the file. The file name to be retrieved is listed in the specification field. The file stored contained all variable names, values, and subfile information. Therefore, the single "GET FILE" card is used in place of all of the necessary control cards. The only necessary cards are, therefore, this one plus a list of task-definition cards and a "FINISH" card.

Many other types of SPSS runs can be made. Optional additions to these runstreams can aid in interpretations and presentations of outputs. These are discussed in the manual. The examples given here do, however, contain the major components necessary for successful card implementation of the program.

III.2. TYPICAL ARTHUR RUNS

INTRODUCTION

An ARTHUR run will be described using the air pollution data base, which will be separated into the four given categories. See Appendix III for a list of the data. The runstream and partial printout is given in Table III.2. There are three versions of ARTHUR in use. The first (1976) was available from Kowalski, the second from Harper, and the most recent from Infometrix, Inc. Data in this and Section III.5 will be concerned with the old version with changes for the new one noted. More information can be found in the manuals.

The program must be assigned and executed according to instructions for the computer facility to be used for the analysis. After that is successfully accomplished, the first card necessary within ARTHUR is a "NEW" card. In ARTHUR, a "$" is used for ending punctuation. This first card is needed to indicate that the program to be run defines a new task or problem. The program prints a portrait of "ARTHUR" with documentation information included. If this is obtained, it can be assumed that the program was accessed correctly. Any errors present must then be in subsequent cards. If the portrait is not obtained, check the methods for accessing the program with the computer facility.

TABLE III.2
ARTHUR Run

Part A Data Listing (1976 version)

```
NEW$
ARTHUR DATA-- MC AND MS ONLY
INPUT,5,10,1,0,10,1,0,1,1$
(I1,2A4,F4.0,2F5.0,2X,6F7.0,F5.0,F6.0)
MN
CU
FE
PB
PART
ORG
SULF
NIT
CL
CO
9999999999999999999999999999999999999999999999999999999999999999
9999999999999999999999999999999999999999999999999999999999999999
SCALE,10,11$
END$
```

Note: in the 1977 and 1980 versions of Arthur, the third card would change to
         INPUT,5,10,10,1,0,1,1,0$

Part B Program Results

```
ARTHUR DATA--    MC AND NS ONLY
NIN      5 INPUT UNIT
NOUT    10 OUTPUT UNIT
NPNT     1
NPCH     0
NVAR    10 NUMBER OF FEATURES READ FROM INPUT FILE
NCAT     1 CATEGORIZED DATA.
NTEST    0 NO TEST SET DATA.
NFMT     1 NUMBER OF RUN-TIME FORMAT CARDS.
NAME     1 FEATURE-NAMES READ FROM CARDS.

YOUR INPUT FORMAT FOR THIS DATA IS...
(I1,2A4,2F5.0,2X,6F7.0,F5.0,F6.0)

YOU HAVE SPECIFIED THE FOLLOWING FEATURE NAMES...
 1 MN
 2 CU
 3 FE
(this list continues on the printout for all features)

TRAINING SET...
MY INDEX  TOTAL MISSING  ***************MISSING MEASUREMENTS*****************
    6        2           0  0  0  0  7  0  9  0
   11        1           0  0  0  0  0  8  0  0
   18        1           0  0  0  0  0  0  0 10
   19        1           0  0  0  0  0  0  0 10
(this list continues for all patterns containing missing values)

HERE-S THE TRAINING SET...
MY  YOUR  NAME     CATEGORY    ***************INPUT FEATURES****************
 1   0   MC75JR12    1.00    4.00-02 2.96+00 1.35+00 2.06+00 7.97+01 1.10+01 2.55+00 8.30-01 1.39+00 1.81+03
 2   0   MC75JR18    1.00    9.00-02 3.11+00 2.63+00 5.61+00 1.69+02 1.81+01 3.06+00 4.39+00 9.33+00 7.30+03
(this list continues for all patterns in the training set)
```

The second card in the deck is a heading card. Anything can be written on it to indicate to the user what procedures and data base are being used. This will be printed at the top of each page in the printout.

INPUT CARD

The third card is the "INPUT" card. This card is used to indicate specifications for the data input. Each specification is separated from the others by commas. These are placed directly after the word "INPUT" and not necessarily in any specific column format as with SPSS. The format for this card, for the 1976 version, is

INPUT,NIN,NOUT,NPNT,NPCH,NVAR,NCAT,NTEST,NFMT,NAME$

where each word after the "INPUT" is to be filled in with a number representing the specification for that word. The choices are

NIN: input unit number—usually "5";

NOUT: output unit number (check with the computer facility for the number of files available for data storage in ARTHUR); in our example it is "10";

NPNT: ≥ 1, input data are to be printed;
   < 1, input data are used but not listed in the printout;

NPCH: ≥ 1, data are to be punched on computer cards (often used to save data after changes have been made);
   < 1, data will not be punched;

NVAR: number of variables to be read from the input file;

NCAT: 1, categorized data are present; this is the choice to be used (even in unsupervised learning techniques) for our applications;

NTEST: 1, test data is present in the data base; test data can be used as "unknowns" to predict classifications after the computer has been trained with the normal data base (see Step II.5);
   < 1, no test set is present;

NFMT: the number of cards that were used to describe each case (or pattern) of data; if it took two cards to list values for all of the variables to be used for each case, then NFMT = 2; in this example, NFMT = 1;

NAME: 1, names for the variables are to be defined by the user;
   < 1, the computer assigns the variables sequence numbers as names.

For our run, this card read

INPUT,5,10,1,0,10,1,0,1,1$

This indicates that data are found in file 5 (cards) and output onto file 10. The data are to be printed back, but not punched. Ten user-defined variables are to be read from the cards, with each case or pattern being present on a single computer card. No test set will be given.

For the 1977 and 1980 versions, the input card has been rearranged to the format

INPUT,NIN,NOUT,NVAR,NCAT,NTEST,NFMT,NAME,IFER$

The only different specification is "IFER" which is used to indicate whether error information is present with each data piece. If it is, this must be listed in the data set immediately following each measurement.

For all versions of ARTHUR, the next card indicates the input format to be used. If the format described in Appendix III is used to create ARTHUR-compatible format, it is listed here. Opening and closing parentheses are used around the format. This is the only information to be placed on this card. Note that the printout in Table III.2 lists this format specification.

If, on the input card, a specification under "NAME" was made to name the variables, they are listed next with each occupying a single physical card. Therefore, as in this case, if "NAME" was "1" and "NVAR" was "10," ten variable names are then given on ten computer cards. These must be followed by a blank card. The printout lists the ten variables back for the user (see the table). These are listed according to the order in which they are encountered in the data base. This is the last control card necessary for ARTHUR.

TASK DEFINITION CARDS

Unlike SPSS, ARTHUR does not list the first task definition card before the data. All procedures are specified after the data deck. Therefore, an indication is made to ARTHUR (after the variable list and blank card) that the data deck follows. Normally the data deck in ARTHUR is bounded on both ends by cards containing 9's in every column. This indicates the points in the deck where the data begin and end. With this procedure, the computer does not need to be told how many cases exist. It will consider all data placed between the 9's cards. If two data format cards were used for each case (NFMT = 2), two cards with 9's in every column are needed before and after the data deck.

At this point on the printout of the old version, the raw data used is listed if NPNT = 1. In the new versions, only the case names and identification are listed. Input is then complete, and the program waits for futher instructions.

After the set(s) of 9's following the data base, task definition cards are listed for the procedures to be used. In ARTHUR, specifications follow the procedure name directly and are separated by commas. Each card ends with a "$."

Note that in Table III.2, the computer was requested to perform a SCALE procedure. The "10,11$" on this card is used to define the chosen options for this procedure. These (and the other options for ARTHUR) will be discussed in Section III.5. After all procedures are defined that are desired, the final ARTHUR card is "END$" to indicate the end of a run. The cards following this are those required to leave the specific computer used for the analysis. Section III.5 lists options available for the given procedures.

TEST DATA

If a set of "test" data is to be included, two changes are necessary. The first change is made on the "INPUT" card. The specification, "NTEST," is set equal to "1." The test data must be added to the data deck directly after the training data and before the specification cards. An additional card (or two if NFMT = 2) of 9's is used between the training set and the test set to indicate the end of the first and the beginning of the second. All other cards remain the same.

When the computer encounters the data, the distinction between the two is recognized. "Training" analyses utilize only the first set. In certain subroutines, the predicted results for the test set are then listed. Comparisons of results from the various subroutines are used to classify "unknowns" into their predicted categories.

In the computer printout in Table III.2, the title "ARTHUR DATA—MC AND NS ONLY" is printed at the top, and the specifications (and their interpretation) from the INPUT card are listed. The input format is shown and the names of the variables listed. A list of "MISSING MEASUREMENTS" is then given. Included in this table are only samples from the first few data pieces. This listing would continue, and would include all patterns that contained missing measurements. Note that out of the first few samples, four (samples numbered 6, 11, 18, and 19) contained missing measurements. Sample 6 had a TOTAL MISSING of two, and the MISSING MEASUREMENTS occurred for variables numbered seven and nine. Sample 11 had one missing value, that for variable number 8. Below this is shown a partial listing of typical patterns that were contained in the original data. The first sample, MC75JR12, belonging to category one, is listed as such. The ten numbers following this are the values for that pattern for the ten input variables. Therefore, the analyses included MN, 0.040; CU, 2.96; FE, 1.35; etc. ARTHUR lists these in scientific notation, even if the original data cards were not. Appendix VIII discusses how the program handles missing data. The estimated value for these would be included in this data listing.

After the data listing is completed, the printout would then show the results from the various statistical procedures.

III.3. TYPICAL BMDP RUNS

INTRODUCTION

In BMDP, each procedure is a program in itself, and not a subroutine within a major program as in SPSS and ARTHUR. Therefore, each program desired must be accessed separately. More than one program cannot be used in a single execution as was shown with SPSS and ARTHUR. However, the control language for BMDP has been devised in such a manner that the same cards with no (or few) changes can be used for all programs.

It will be assumed that the user has information on how to access the individual computer programs for BMDP (see the computer consultant for your facility). The example given here will use BMDP6D for two-dimensional variable plots on the air pollution data.

BMDP utilizes an English-based control language. Instructions are written in sentences which are grouped into paragraphs. For most analyses, three general control card paragraphs are used: the problem, input, and variable paragraphs. These are similar for all programs. Paragraphs then follow which describe the detailed specifications for the given problem. The paragraph name is listed, preceded by a slash and followed by sentences describing the paragraph. Sentences end with periods and values or names in the sentences are separated by some type of punctuation (slashes, commas, periods, or equal signs) or blanks. No specific format for the control cards is necessary. Blanks can be used or skips to new cards made at any time. Paragraphs can continue on the same card (although it is easier to read if each paragraph is given on a separate one).

One type of sentence to be used is a general assignment sentence where an item is assigned a given value or values. The words "IS" or "ARE" can be used in place of the equal sign "=."

In most BMDP paragraphs, preassigned values exist for many of the specifications and are assumed to be valid unless other values are specified. Sentences within each paragraph can be arranged in any order.

CARD DECK

The first paragraph is the "/PROBLEM" paragraph. This can be abbreviated as "/PROB." No other specifications are necessary, but can be helpful. A sentence listing the title is useful. This will be given at the beginning of the printout to indicate the problem under consideration. Table III.3 gives an example for the air pollution data. The title may contain up to 160 characters. It must be enclosed in apostrophes if any blanks or special characters are present. The title sentence thus becomes

TITLE = '456 CASES- FE VS. PARTICULATES'.

The word "IS" could have been used in place of the equal sign. The slash can be placed at the end of the paragraph. However, it is easier to place it directly before the next paragraph name so as not to forget it. Either method is equally valid.

The second paragraph needed is an "/INPUT" paragraph. This paragraph requires specifications for the number of variables and the format of the variables. Each of these is a sentence in the paragraph. The number of cases can be (but need not be) specified. For our example, ten variables were used. In BMDP, the two case names are considered as two additional variables. Therefore, the sentence is given as "VARIABLE

TABLE III.3
BMDP6D Computer Run

Part A   Data Listing

```
/PROBLEM      TITLE='456 CASES-  FE VS. PARTICULATES'.
/INPUT   VARIABLE = 12.   FORMAT = '(1X,2A4,4X,2F5.0,2X,6F7.0,F5.0,F6.0)'.
         CASES = 456.
/VARIABLE    NAME =CASE1,CASE2,MN,CU,FE,PB,PART,ORG,SULF,NIT,CL,CO.
         LABEL ARE CASE1,CASE2.
         MISSING = (3)99999,99999,6*9999999,99999,999999.
/PLOT        XVAR=FE.  YVAR=PART.
/END

    data cards are  placed here

/FINISH
```


Part B   First Page of Output

```
   PROGRAM CONTROL INFORMATION
         /PROBLEM      TITLE='456 CASES-  FE VS. PARTICULATES'.
         /INPUT   VARIABLE = 12.   FORMAT = '(1X,2A4,4X,2F5.0,2X,6F7.0,F5.0,F6.0)'.
             CASES = 456.
         /VARIABLE    NAME =CASE1,CASE2,MN,CU,FE,PB,PART,ORG,SULF,NIT,CL,CO.
             LABEL ARE CASE1,CASE2.
             MISSING = (3)99999,99999,6*9999999,99999,999999.
         /PLOT        XVAR=FE.  YVAR=PART.
         /END

   PROBLEM TITLE . . . . . . .456 CASES- FE VS. PARTICULATES

   NUMBER OF VARIABLES TO READ IN. . . . . . . .      12
   NUMBER OF VARIABLES ADDED BE TRANSFORMATIONS. .     0
   TOTAL NUMBER OF VARIABLES . . . . . . . . . .      12
   NUMBER OF CASES  TO READ IN . . . . . . . . .     456
   CASE LABELING VARIABLES . . . . . . . . . . .    CASE1    CASE2
   LIMITS AND  MISSING VALUES CHECKED BEFORE TRANSFORMATIONS
   BLANKS ARE. . . . . . . . . . . . . . . . .     ZEROS
   INPUT UNIT NUMBER . . . . . . . . . . . . . .       5
   REWIND INPUT UNIT PRIOR TO READING. . DATA. . .     NO

   INPUT FORMAT
         (1X,2A4,4X,2F5.0,2X,6F7.0,F5.0,F6.0)

   VARIABLES TO BE USED
         3  MN              4  CU           5  FE          6  PB          7  PAR
         8  ORG             9  SULF        10  NIT        11  CL         12  CO
   NUMBER OF CASES READ. . . . . . . . . . . .     456
```

= 12." The format for these is similar to that used in ARTHUR. However, in this case, no ID number is required. The format is enclosed in apostrophes and parentheses:

$$\text{FORMAT} = \text{'(1X,2A4,4X,2F5.0,2X,6F7.0,F5.0,F6.0)'}.$$

Since, in this problem, only a bivariable plot of FE versus PART is desired, the variable number could be given as four (two case names, FE and PART) and the format changed to reflect this. Other variables would be crossed out of the variable format. However, it is easier to keep the variable list and format specifications as general as possible. These cards can then be used for other BMDP programs as well. The variables to be chosen from the list for a specific problem can later be designated.

Since BMDP does not always print back the data input, it is suggested that a case sentence be used, although it is not necessary. This will make sure that the computer is reading the number of cases that the user knows exists. This card is given as "CASES = 456." These three sentences are the typical ones to be included in most "INPUT" paragraphs for card deck inputs.

VARIABLE INFORMATION

The third required paragraph is the "/VARIABLE" paragraph. This paragraph is optional but is usually used to further describe the data. One sentence present is the name sentence, used to name the variables in the list. The names must be ordered in the same manner as the data in the data deck. Names are restricted to eight characters. Be sure to name the case names as well. Twelve names should be present to match the twelve variables in the list. This sentence then becomes

NAME = CASE1,CASE2,MN,CU,FE,PB,PART,ORG,SULF,NIT,CL,CO.

If the case names are to be used as labels for the cases, one or two variables can be used for this. Each may not exceed four letters and must be read under A-format. These must be identified as such with a label sentence such as

LABEL ARE CASE1,CASE2.

Missing values are also designated in this paragraph. If blank values are to be considered as missing data, a sentence can be used to designate this:

BLANK = MISSING.

These are then eliminated from all analyses. If other missing value codes are used (as with ARTHUR-compatible formats) these must be indicated. In this case the sentence is

MISSING = (3)99999,99999,6*9999999,99999,999999.

The three in parenthesis is used as a tab function. The first two variables are skipped, and the third one is assigned 99999 as a missing code. Five nines are used since MN was assigned a format of F5.0. The next variable, CU, also used five spaces. The next six variables each took seven spaces. Either six sets of seven nines could be used or a 6*9999999. This second convention is used to indicate that the value is to be considered six times. The last two variables used five and six columns, respectively, indicating formats of F5.0 and F6.0.

If a subset of variables is to be analyzed, it may be specified at this point in a "USE" sentence. Often this is available in the specific program specifications as well. Since the variables to be used vary with the analysis, it is better to specify them in the individual program so that this variable paragraph remains as generally applicable as possible.

Maximums and minimums can also be specified for each variable. For instance, if the sixth variable had a maximum value limit of 50, the sentence would read

$$\text{MAX} = (6)50.$$

Many specifications may be made for maximum and minimum values. For example, assume that the upper acceptable limits for the fourth through seventh variables were 40, 22.3, 0.32, and 35.0 respectively, and the minimum values for the fourth, sixth, and ninth variables were 2.0, 3.0, and 1.0 respectively. Two sentences in the variable paragraph would be necessary to define these. The sentences could be inserted anywhere in the variable paragraph, and would take the from

$$\text{MAX} = (4)40.,22.3,0.32,35.0. \quad \text{MIN} = (4)2.0,(6)3.0,(9)1.0.$$

DATA GROUPING

If the data are to be stratified into groups, the next paragraph would be the "/GROUP" paragraph. Note that this is not the "/CATEGORY" paragraph as might be expected. The category paragraph, in BMDP, is used to group variable values for frequency tables. Therefore, the "/GROUP" paragraph is used to stratify data. BMDP has an option for omitting this if ARTHUR-type data are analyzed. If the codes for the grouping variable consist of less than ten values, there is no need for a group paragraph unless you want to specify names for the groups. In our case, the values 1.0, 2.0, 3.0, and 4.0 were used to indicate the categories MC, MS, NS, and SC, respectively. There are only four codes, and therefore a grouping paragraph was not necessary. However, for those analyses where groups are to be considered, a variable must be added to the list to define the group variable and the format statement must be adjusted appropriately. The name would also have to be added to the variable names list. This would change the input and variable paragraphs to

```
/INPUT      VARIABLES = 13.
            FORMAT = '(1X,2A4,F4.0,2F5.0,2X,6F7.0,F5.0,F6.0)'.
            CASES = 456.
/VARIABLE   NAME = CASE1,CASE2,GROUP,MN,CU,FE,PB,PART,
            ORG,SULF,NIT,CL,CO.
            LABEL ARE CASE1,CASE2.
            MISSING = (4)99999,99999,6*9999999,99999,999999.
```

If more than ten distinct values exist for the grouping variable, a "/GROUP" paragraph is required, and is placed after the "/VARIABLE" and before the "/END" one to designate group definitions. If the grouping variable contains a finite number of values (or codes), but only a subset of these values are desired category definitions, a "CODE" sentence is necessary. This takes the format

$$\text{CODE (A)} = \#_1, \#_2, \ldots$$

where A is the subscript number of the variable to be used as the grouping one, and $\#_1$, $\#_2$, etc., are the values for that variable to be used to designate category membership. Up to ten may be specified. Any case where the grouping variable takes on any other value not given in this list will not be included in the analysis. The grouping variable can alternately be split into categories by using a "CUTPOINT" sentence with the format

$$\text{CUTPOINT (A)} = \#_1, \#_2, \ldots$$

where A is the subscript number of the variable to be used as the grouping one. The values listed as $\#_1$, $\#_2$, $\ldots$ are the upper limits of the intervals used to divide this variable into categories. If desired, the "/GROUP" paragraph can also be used to name the categories. The individual procedures would indicate whether the groups are to be considered as a whole or individually.

The next paragraph is either one specified for the given program to indicate desired options and methods, or an "/END" paragraph to indicate the end of the control cards. No other specifications for it are required. The data deck then follows the end paragraph directly. BMDP programs dated before August 1977 require a "/FINISH" card following the data deck to signify the end of the BMDP procedures before the specification for the system to terminate. Those programs dated after this do not require one, and end only with the specific computer system's finish card after the data.

On the printout, the control cards are listed in the given order. A summary of the input is given below this. This should be carefully checked to see if it is correct. Results are then plotted for the statistical program. Specific examples of options for the BMDP programs are given in Section III.6.

III.4. SPSS IMPLEMENTATIONS

INTRODUCTION

In SPSS, the first set of cards (described in III.1) is used to define the data base structure and the variables to be considered. Directly following these in the card deck is a set of cards used to define the first statistical procedure that is to be performed. This set is referred to as the task definition cards. A "READ INPUT DATA" card is needed next, directly followed by the data deck. After the last data card, any number of task definition card sets may be utilized in order to perform further statistical procedures. After the last one, a "FINISH" card is included followed by those cards normally needed to exit the computer system. More details with regard to the task definition set of cards for SPSS will be given in this section.

Three types of cards may be included in this set. The first (which is required) is the task definition card itself. The first 15 columns of the card are used for the name of the statistical procedure to be performed. The choices include those given in Appendix IV and are referred to throughout this book by setting the name in small capital letters. This part of the task definition card will be referred to as the control word *(CW)*. The other half of this card (columns 16–80) is used to define the problem to be performed and is referred to as the specifications field *(SP)*. This may continue on subsequent cards, if necessary, but columns 1–15 must be left blank on the other cards.

The second type of card in a task definition set is the options card (which itself is optional). This card provides further information for the analysis. On this card, the word "OPTIONS" is placed in the first seven columns. Beginning in column 16, a listing of option numbers relating to those options to be performed is made. The list is subroutine specific and is considered below for each under *OP*.

The third card (also optional in the set) is a statistics card which selects additional statistics for the analysis to be performed. The word "STATISTICS" is listed in columns 1–10. Beginning in column 16, a listing of numbers, relating to those statistics chosen for inclusion, is made. Again, these choices are subroutine-specific and will be given below under *ST*. The "OPTIONS" and "STATISTICS" cards may be included in the deck with either given first.

In this section, each of the major subroutines from SPSS utilized in the Chapter II discussions will be considered individually. A summary of some of the choices for the task definition card sets will be given. The ones found most commonly useful for the authors' analyses will be shown, although many other choices are available. The listing here, therefore, is chosen to reflect those useful for most routine analyses. Examples are given under "EXAMPLE." The numbers below this in each represent the card columns for the entry. Spacing is not critical; however, the SP must not begin before column 16. The reader is encouraged to study other examples and choices in the SPSS manual.

CANCORR

CANCORR calculates the correlations between two sets of variables, each of which has meaning as a set.

CW: CANCORR

SP: The first specification necessary is a "VARIABLES =" in columns 16–25 followed by a list of variables to be included in the analyses. The word "TO" may be used for adjacent variables in the data base. This specification is followed by a slash. The next required information is a "RELATE =" parameter followed by the two lists of variables to be considered as the sets.

The lists are separated by the word "WITH." The set before the "WITH" is then related to those variables after it. A slash after the second set may be followed by another "RELATE =" specification.

OP: None needed.

ST: 1, prints out means and standard deviations of the individual variables.

2, Outputs the simple correlation coefficients.

Example:

```
1234567890123456789012345678901234567890123456789012345678901234567890
a. CANCORR       VARIABLES =ITM1 TO ITM6/RELATE =ITM1 TO ITM3 WITH ITM4,ITM5,ITM6/
b. CANCORR       VARIABLES =ITM1 to ITM4,ITM6/RELATE =ITM1,ITM2,WITH ITM3,ITM4/RELATE =ITM1 TO ITM3
                 WITH ITM4,ITM6/
```

Example (a) will relate ITM1, ITM2, and ITM3 as a set to ITM4, ITM5, and ITM6 as the second set. Example (b) considers ITM1, ITM2, ITM3, ITM4, and ITM6. The first problem relates ITM1 and ITM2 with ITM3 and ITM4. The second problem relates ITM1, ITM2, and ITM3 with ITM4 and ITM6.

CONDESCRIPTIVE

CONDESCRIPTIVE is used for introductory univariate statistics.

CW: CONDESCRIPTIVE

SP: A list of variables to be included for the statistical manipulations. The word "TO" may be used. If all variables are to be considered, the word ALL is used in the specifications field.

OP: None usually used.

ST: ALL is used, which causes a mean, standard error, standard deviation, variance, kurtosis, skewness, range, minimum, and maximum value to be listed for each variable.

Example:

```
1234567890123456789012345678901234567890
a. CONDESCRIPTIVE VAR1,VAR2,VAR3
b. CONDESCRIPTIVE VAR1 TO VAR3
c. CONDESCRIPTIVE ALL
```

Examples (a) and (b) will calculate the statistics chosen on variables VAR1, VAR2, and VAR3. Example (c) will do the same for all variables in the data set.

CROSSTABS

CROSSTABS calculates joint frequency distribution of cases.

CW: CROSSTABS

SP: May be used with either the integer mode or the general mode. The integer mode can utilize only integer numbers but has the advantage of being faster and having more statistical measures of association. Any type of variable having less than 250 unique values can be entered under the general mode. The integer mode needs two specifications, a "VARIABLES =" and "TABLES =." Only the second is necessary in the general mode. Both "TABLES =" specifications are identical. "TABLES =" is written in columns 16–22 followed by the variable list for the tables to be created. Each variable name is separated by the word "BY." Multiple tables can be requested by placing more than one variable name to the right or left of the word "BY." The "TO" convention may be used. This will produce a table for each variable to the left of the "BY" for each variable to the right. More than one list may follow the "TABLES =," each being separated by a slash. Two-way to ten-way multiple tables may also be produced. An n-way table must contain n variables, each separated by the word "BY" (see example below). In the integer mode, a "VARIABLES =" specification must precede the "TABLES =." "VARIABLES =" occupies columns 16–25 followed by the variables to be included. Each variable listed is followed by the lowest and highest value for it in parentheses. If more than one variable has the same value limitations, these variables can be listed together followed by their values in parentheses. A slash ends the "VARIABLES =" specification followed directly by the "TABLES =" specification, utilizing a format similar to the one given above for the general mode.

OP: None used.

ST: Causes the following if individual numbers are used: (1) chi-square, (2) phi for 2×2 tables or Cramer's V for larger, (3) contingency coefficients, (4) lambda, (5) uncertainty coefficients, (6) Kendall's tau b, (7) Kendall's tau c, (8) gamma, (9) Somers' D, (10) eta.

Example:

```
12345678901234567890123456789012345678901234567890123456789012345678901234567890
a. CROSSTABS        TABLES =VAR1 BY VAR2
b. CROSSTABS        TABLES =VAR1 BY VAR2 BY VAR3
c. CROSSTABS        TABLES =VAR1,VAR2 BY VAR3
d. CROSSTABS        TABLES =VAR1 BY VAR3/VAR2 BY VAR3
e. CROSSTABS        TABLES =VAR1 TO VAR2 BY VAR3
f. CROSSTABS        VARIABLES =VAR1(0,10),VAR2,VAR3(0,15),VAR4(0,20)/TABLES =VAR1
                    TO VAR3 BY VAR4
```

Example (a) produces a table with the values for VAR1 as the row variable and VAR2 as the column variable. The table will list the number of cases contained in each cell of the table. Example (b) produces multiple tables. Each considers a single value for VAR3 and produces a VAR1 by VAR2 table. Therefore, the number of tables produces equals the number of unique values for VAR3. Examples (c), (d), and (e) show three ways to indicate the problem of creating the tables for both VAR1 by VAR3 and VAR2 by VAR3. Example (f) utilizes the integer mode. VAR1 takes on values between zero and ten; VAR2 and VAR3 between zero and 15; and VAR4 between zero and 20. Three tables are producers—VAR1 with VAR4, VAR2 with VAR4, and VAR3 with VAR4. Note how the specification was continued on a second card (as necessary) beginning in the 16th column.

DISCRIMINANT

DISCRIMINANT—discriminant analyses are performed for class separation.

CW: DISCRIMINANT

SP: Three specifications are necessary for each procedure. The first is "GROUPS =" followed by the name of the variable used to define the groups or classes. A minimum and maximum value to be considered for that variable may be included in parentheses following the variable name. The variable must be integer valued. Alternatively, a "GROUPS = SUBFILES" specification may be made which will consider each subfile from the original data deck as a group. In this procedure, no cases are considered unclassified. In the first method, using a variable to define the groups, any cases with values for that variable not being an integer between the minimum and maximum values specified will be considered unclassified and will later be used in the classification phase of the problem. The "GROUPS =" specification is ended with a slash followed by "VARI-ABLES =," a variable list to be considered for the problem, and a second slash. The third specification (which may optionally be repeated as many times as needed) is an "ANALYSIS =" specification, which defines the exact problem to be performed. After "ANALYSIS =" is listed the variables for that specific problem followed optionally by an inclusion level in parentheses. These variables must have been listed in the "VARI-ABLES =" section of the card. Inclusion levels are listed for stepwise procedures only. The variables with the highest value will be entered before those with lower values in a stepwise manner. If the values are even, all variables at that level are entered simultaneously. If the inclusion values are odd, those variables with the same inclusion value are entered according to some entry criteria. After the "ANALYSIS =" specification is a slash. A "METHODS =" specification can then be made to further define the problem. The method here determines under what criteria variables are to be entered into the stepwise problem. The choices are

a. "METHOD = DIRECT": No stepwise procedure is used. All variables are considered simultaneously.
b. "METHOD = WILKS": Smallest Wilks' lambda value.
c. "METHOD = MAHAL": Largest Mahalanobis distance between the two closest groups.
d. "METHOD = MINMAXF": Largest pairwise F.
e. "METHOD = MINRESID": Smallest unexplained variance.
f. "METHOD = RAO": largest increase in Rao's V.

A slash follows the "METHOD =" specification. This may optionally be followed by a second and subsequent "ANALYSIS =" specification for the problem using the same "VARIABLES =" and "GROUPS =" definition.

OP: 5, Prints classification results table.

6, Prints discriminant scores and classification information.

7, Single plot containing all cases.

8, Individual plots containing the cases for each group.

11, Prints unstandardized discriminant function coefficients.

12, Prints classification functions.

13, VARIMAX rotation of the discriminant functions.

ST: 1, Means of the discriminating variables.

2, Standard deviations of the discriminating variables.

Example:

```
1234567890123456789012345678901234567890123456789012345678901234567890
a. DISCRIMINANT    GROUPS =SUBFILES/VARIABLES =VAR1 TO VAR9/ANALYSIS =VAR1 TO VAR5/ANALYSIS =VAR2
                   TO VAR8/METHOD =RAO/
b. DISCRIMINANT    GROUPS =VAR1(1,5)/VARIABLES =VAR2,VAR3,VAR4,VAR5/ANALYSIS =VAR2,
                   VAR3(3),VAR4,VAR5(2)/METHOD =WILKS/
```

Example (a) utilizes the subfile separation as the definition of groups. The variables to be included are VAR1 to VAR9. The first analysis utilized variables VAR1 to VAR5. No method for entry was specified. The default method is direct. The second analysis in this problem utilized VAR2 to VAR8. No inclusion level is given so it is assumed to be "1." Therefore the variables are entered stepwise in the order determined by their effect on the Rao's V statistic. Example (b) used VAR1 as the grouping variable. Values of 1, 2, 3, 4, and 5 for VAR1 constitute the group separation. Any cases taking on values other than this are not included in the classification scheme, but are reserved for the final predictions. The variables to be included in the discrimination procedure are VAR2, VAR3, VAR4, and VAR5. The analysis procedure first utilizes VAR2 and VAR3 stepwise according to Wilks' lambda values and then VAR4 and VAR5 are entered simultaneously into the problem.

FACTOR

FACTOR performs a factor analysis on the data.

CW: FACTOR

SP: The first necessary specification is "VARIABLES = " in columns 16–25 followed by a list of the variables to be used for the problem. The "TO" convention may be specified. The last variable name is followed by a slash. The second specification needed is a "TYPE = " to select the method for choosing the initial factors. Most commonly, either "TYPE = PA1" (principal factoring without iterations) or "TYPE = PA2" (principal factoring with iterations) is chosen. PA2 is the default method. This specification is followed by a slash. The main diagonals of the correlation matrix may be altered at this point using a "DIAGONAL = " specification. See manual for this usage. The number of factors extracted or the number of permitted iterations may also be specified. The next specification commonly used is the "ROTATE = " with the choices of "VARIMAX," "QUARTIMAX," "EQUIMAX", "OBLIQUE", or "NOROTATE." See Step II.5c for an explanation of this. The default method is "VARIMAX." The rotate specification is followed by a slash. A second "VARIABLES = " specification can then be made to begin a second problem.

OP: None needed.

ST: 2, Prints simple zero order correlation matrix.
5, Prints the initial unrotated factor matrix.
8, Plots the rotated factors.

Example:

```
1234567890123456789012345678901234567890123456789012345678901234567890
a. FACTOR          VARIABLES =VAR1 TO VAR6/TYPE =PA1/ROTATE =VARIMAX/VARIABLES =VAR1 TO VAR6/ROTATE =
                   EQUIMAX/
b. FACTOR          VARIABLES =VAR1,VAR2,VAR3/
```

The first example considers VAR1 to VAR6 utilizing a PA1 initial factor extraction with no iterations. A VARIMAX rotation of these factors is performed. The second problem in this example considers the same variables utilizing a PA2 initial factor extraction (by default) and an EQUIMAX rotation. Example (b) uses VAR1, VAR2, VAR3, initially extracting factors by PA2, and rotating these according to EQUIMAX (both by default).

FREQUENCIES

FREQUENCIES is a statistical procedure like CONDESCRIPTIVE but for noncontinuous data.

CW: FREQUENCIES

SP: Must list either "GENERAL = " or "INTEGER = ." The first will give frequency tables for all types of variables. The second is used for those variables whose values take on only integer values. The "INTEGER = " is followed by a list of those variables to be included in the analysis. The word "TO" may be used. If all variables are to be included, the word "ALL" can be used. If the "INTEGER = " mode is used, the low and high values for each variable are listed in parentheses following the variable name.

OP: 8, Requests a histogram plot for each variable.

ST: ALL is used which causes the mean, standard error, median, mode, standard deviation, variance, kurtosis, skewness, range, minimum, and maximum values to be calculated for each variable.

Example:

```
1234567890123456789012345678901234567890123456789012345678901234567890
a. FREQUENCIES     GENERAL =VAR1,VAR2,VAR3
b. FREQUENCIES     GENERAL =ALL
c. FREQUENCIES     INTEGER =VAR1(0,5),VAR2 TO VAR3(0,6)
d. FREQUENCIES     GENERAL =ALL
   OPTIONS         8
   STATISTICS      ALL
```

Examples (a) and (b) request that all values be considered for the analysis. Example (a) considers VAR1, VAR2, and VAR3. Example (b) uses all of the variables contained in the data set. Example (c) considers integer values for VAR1 from zero to five and for VAR2 and VAR3 from zero to six. Example (d) requests that all values for all variables in the data set are to be used. Histograms are plotted for each variable, and all of the various statistics are calculated and listed.

NONPAR CORR

NONPAR CORR calculates nonparametric correlation coefficients (see Appendix VII).

CW: NONPAR CORR
SP: Contains a list or lists of variables to be considered. The word "TO" may be used. The word "WITH" is used to pair the variable lists. Each variable before the "WITH" is correlated with each variable after. Slashes may be used to separate individual correlation problems.
OP: 3, Two-tailed significance tests are applied instead of one-tailed tests.
     6, Both Kendall and Spearman correlations are calculated (the default is to calculate only Spearman values).
ST: None available.

Example:
```
1234567890123456789012345678901234567890123456789 0
a.  NONPAR CORR      ITM1 WITH ITM2
b.  NONPAR CORR      ITM1 TO ITM3 WITH ITM4 TO ITM6
c.  NONPAR CORR      ITM1 TO ITM3
d.  NONPAR CORR      ITM1,ITM2/ITM3 TO ITM5
    OPTIONS          3,6
```

Example (a) will calculate the correlation for ITM1 with ITM2. Example (b) will calculate correlations for ITM1, ITM2, and ITM3 with ITM4, ITM5, and ITM6. Example (c) will calculate correlations for ITM1, ITM2, and ITM3 with ITM1, ITM2, and ITM3. Example (d) calculates ITM1 and ITM2 with ITM1 and ITM2 and then calculates ITM3, ITM4, and ITM5 with ITM3, ITM4, and ITM5. Two-tail significance testing is done and both Kendall and Spearman coefficients are calculated.

PARTIAL CORR

PARTIAL CORR calculates the correlation between two variables while controlling for the effects of one (or more) other variables.

CW: PARTIAL CORR
SP: A correlation list must be given. The word "WITH" is used to separate the two groups. All variables before the "WITH" are correlated with each variable after. The word "BY" is then used, followed by the variables used as controls in the analyses. An order value is given in parentheses to specify

how the controlling is to be performed. If a "(1)" is used, each control variable is applied individually. If a "(2)" is used, the control variables are considered two at a time. A "(1,2)" considers each individually and then two at a time. This may continue up to an order of five.

OP: 3, Two-tail test of significance.

ST: 1, Causes zero-order correlations also to be listed.

2, Lists means and standard deviations for the variables.

Example:

```
1234567890123456789012345678901234567890123456789012345
a. PARTIAL CORR    VAR1 WITH VAR2 BY VAR3,VAR4(2)
b. PARTIAL CORR    VAR1,VAR2 WITH VAR3 BY VAR4,VAR5(1,2)
c. PARTIAL CORR    VAR1 WITH VAR3 TO VAR5 BY VAR6(1)
```

In example (a) the correlation of VAR1 with VAR2 is calculated controlling for both VAR3 and VAR4 simultaneously. Example (b) calculates VAR1 with VAR3 and VAR2 with VAR3, both analyses controlling first for VAR4 and VAR5 individually and then simultaneously. Example (c) calculates the correlations of VAR1 with VAR3, VAR1 with VAR4, and VAR1 with VAR5, all controlling for VAR6.

PEARSON CORR

PEARSON CORR calculates the biviariate Pearson product–moment correlations.

CW: PEARSON CORR

SP: The variable names (or lists) for which correlations are to be calculated. The word "WITH" is used to pair the variable lists. All variables before the "WITH" are correlated with each variable after. Slashes may be used to separate the first correlation problem from subsequent ones.

OP: 3, Two-tailed significance tests are applied instead of one-tailed tests.

ST: 1, Causes means and standard deviations for each variable to be printed.

Example:

```
123456789012345678901234567890123456789012345678901234567890
a. PEARSON CORR    ITM WITH ITM2
b. PEARSON CORR    ITM1 TO ITM3 WITH ITM4 to ITM6
c. PEARSON CORR    ITM1,ITM2, ITM3, ITM4
d. PEARSON CORR    ITM1 WITH ITM2/ITM3 WITH ITM4
```

Example (a) calculates the correlation of ITM1 with ITM2. Example (b) considered ITM1 with ITM4, ITM1 with ITM5, ITM1 with ITM6, ITM2 with ITM4, ITM2 with ITM5, etc. Example (c) calculates all possible two-member subsets of the four items (ITM1 with ITM2, ITM3, and ITM4 individually, etc.). Example (d) calculates ITM1 with ITM2 and then ITM3 with ITM4.

REGRESSION

REGRESSION performs multiple regression analyses with an optional stepwise inclusion of variables.

CW: REGRESSION

SP: The first necessary specification is a list of variables to be considered for the problem. "VARIABLES = " is given in columns 16–25 followed by the variable list. The "TO" convention may be used. A slash ends the list followed by "REGRESSION = " and a design statement. The design statement indicates the problem to be considered. The list includes the dependent variable, the word "WITH," and then the set of independent variables to be considered. Following this, in parentheses, is an inclusion level indicator. If the parentheses enclose a single even number, the problem will proceed as a simple multiple regression analysis where all independent variables are considered simultaneously. If certain independent variables are to be considered previous to others in the set, a hierarchy of variables can be indicated. The larger even numbers (in parentheses after certain variables) indicate that these are to be considered in the order of their values, with the higher values being considered first. Those variables with the same value of inclusion will be considered simultaneously. A stepwise procedure may also be indicated by using odd inclusion level values. The computer enters variables stepwise according to some statistical criteria of usefulness in explaining the variance in the problem. Hierarchy of sets of variables for the stepwise procedure can be designated by using larger odd inclusion values for those variables to be considered first. A second "REGRESSION = " problem may then be defined after the first one and a slash.

OP: None.

ST: 1, The correlation matrix is printed.

2, Means and standard deviations of the variables are listed.

6, Standardized residuals are plotted against the standardized predicted values.

Example:

```
123456789012345678901234567890123456789012345678901234567890123456789012345
a. REGRESSION      VARIABLES =VAR1 TO VAR6/REGRESSION =VAR1 WITH VAR2 TO VAR6(2)/REGRESSION =VAR1
                   WITH VAR2 TO VAR6(3)/
b. REGRESSION      VARIABLES =VAR1,VAR2,VAR3,VAR4,VAR5,VAR6/REGRESSION =VAR1 WITH
                   VAR2,VAR3(4),VAR4,VAR5(2)/REGRESSION =VAR1 WITH VAR2,VAR3(3),VAR4,VAR5(1)/
```

Examples (a) and (b) consider all variables between VAR1 and VAR6. Both regressions in example (a) utilize VAR1 as the dependent variable. The first regression uses VAR2, VAR3, VAR4, VAR5, and VAR6 as independent variables and these are entered into the equation simultaneously. The second regression in example (a) enters these same variables according to the statistical criteria of usefulness. Example (b) considers the same variables. Both regression problems in this example again utilize VAR1 as the dependent variable. The first considers VAR2 and VAR3 simultaneously followed by VAR4 and VAR5 simultaneously for the independent variables. The second regression considers VAR2 and VAR3 stepwise first followed by VAR4 and VAR5 stepwise.

SCATTERGRAM

SCATTERGRAM prints two-dimensional variable versus variable plots.

CW: SCATTERGRAM

SP: The variable list for the desired plots. The word "WITH" is used in a manner similar to PEARSON CORR. After each variable, the lowest and highest values to be considered may be listed in parentheses.

OP: 2, Missing values are excluded from the graph and from the statistics listwise.

ST: ALL will give information on the simple $Y = mX + b$ regression line and the simple Pearson product–moment correlation coefficient.

Example:
```
123456789012345678901234567890123456789012345678901234567890
a. SCATTERGRAM    VAR1 TO VAR3
b. SCATTERGRAM    VAR1 TO VAR3 WITH VAR4
c. SCATTERGRAM    VAR1(0,15),VAR2(3,HIGHEST)/VAR3(LOWEST,30) WITH VAR4
```

Example (a) will plot VAR1 with VAR2, VAR1 with VAR3, and VAR2 with VAR3 (all possible combinations). Example (b) will plot VAR1 with VAR4, VAR2 with VAR4, and VAR3 with VAR4. Example (c) will plot all values for VAR1 between zero and 15 with VAR2, only including those values for VAR2 above three. It will then plot VAR3, including any values below thirty with VAR4 (including all values for VAR4).

T-TEST

T-TEST is a test for differences in two sample means.

CW: T-TEST

SP: Must define the groups to be compared, and the variables to consider in those groups. This must begin with "GROUPS =" followed by the group definitions. This can be accomplished by listing the grouping variable followed by (in parentheses) the value of that variable used for separation. All values greater than or equal to this are considered in group 1. All others are group 2 cases. A slash ends this portion of the specification. Alternatively, a "GROUPS = n_1,n_2/" may be used to consider the first n_1 cases as group one and the next n_2 as group 2. In either case, the slash is followed by "VARIABLE =" and a list of variables to be included in the analysis.

OP: None needed.

ST: None available.

Example:
```
123456789012345678901234567890123456789012345678901234567890012345
a. T-TEST         GROUPS =CAT(3)/VARIABLES =ITM1 TO ITM3
b. T-TEST         GROUPS =10,25/VARIABLES =VAR1,VAR2
```

Example (a) will consider all cases with a value for variable CAT greater than or equal to 3 as the first group. Those with values less than 3 form the second group. Analyses are done to compare the means of ITM1, ITM2, and ITM3 individually. Example (b) forms its groups by considering the first ten cases as the first group and the next 25 as the second group. VAR1 and VAR2 will be considered.

(This program can also be used to test the difference in the means of two variables with each being analyzed for all cases in the data—see manual for this usage.)

III.5. ARTHUR IMPLEMENTATIONS

INTRODUCTION

In ARTHUR, each program is a subroutine, and therefore, the introductory cards need only be given once to perform a variety of techniques. These cards are described in Section III.2. ARTHUR contains very few options compared to BMDP and SPSS. For those available, the program has default options for the most commonly used ones. Therefore, typically, the subroutines are called, and the default methods are used.

The new versions (1977 and 1980) of ARTHUR expand the number of available options. Subroutines CHANGE and KARLOV from the old version have been broken down further into other subroutines to expand the choices available.

In this discussion, we will show the options commonly used. The examples given will describe most of the default parameters. See the manual for further details.

Common to most of the ARTHUR subroutines are the specifications NIN (number of the input unit for the data matrix), NOUT (number of the output unit for the subroutine), NPNT (a choice of whether the data matrix is to be printed), and NPCH (a choice of whether the output data matrix is to be punched). Prerequisites for the NIN data are given in the discussions for each subroutine.

CORREL

CORREL generates Pearson product correlation coefficients for all variable pairs.

Format: CORREL,NIN,NPNT,TS$ (1976,1977)
  NPNT: ≤ 0 correlation matrix printed
    >0 covariance and correlation matrices printed
  TS: ≤ 0 Student's t assumed to be 2.0
    >0 Student's t is value given for TS
Example: CORREL,10$

Input unit for the data is number 10. The correlation matrix (but not the covariance matrix) is printed utilizing the Student's t value of 2.0.

DISTANCE

With DISTANCE the distance matrix between the n-dimensional points is generated.

Format: DISTANCE,NIN,NOUT,NPNT,N,XLF,XHF $ (1976)

DISTANCE,NIN,NOUT,NTYPE,N,XLF,XHF $ (1977)

NTYPE (1977): $=0$ Mahalonobis distance

$=1$ city block distance

$=2$ ratio distance

$=3$ Behrens–Fisher test

$=4$ overlap integral

$=5$ Gaussian area

$=6$ error weight city block distance

N(1976): <0 ratio distance

$=0$ Euclidean distance

$=1$ city block distance

>1 Mahalonobis distance of order N

N(1977): ≤ 2 distance exponent equals 2

>2 distance exponent equals N

Example: DISTANCE,10$

The data matrix from NIN equals 10 is used and the distance matrix is output to NOUT. In both the 1976 and 1977 versions, this default utilizes a Mahalanobis distance with an exponent of 2.

HIER

HIER generates and prints a hierarchical clustering dendrogram for the data as an unsupervised learning technique.

Format: HIER,NIN,IWAIT$ (1976)

HIER,NIN,NPNT,ITYPE,IPAGE   (1977)

IWAIT: ≤ 0 each pattern is given equal weight in determining linkage levels

>0 each category is given equal weight in determining linkage levels

NPNT: <0 histogram plotted only

>0 histogram, cluster linkage levels, and members listed

ITYPE: <0 single link cluster

>0 complete linkage clustering

IPAGE: ≤ 1 dendrogram printed on one page

$=2$ dendrogram printed on two pages

≥ 3 dendrogram printed on three pages

Example: HIER,12$

Distance data must be used as the input data. This example reads this from unit number 12. In the 1976 version, each pattern is given equal weight in the process. In the 1977 version, histograms are plotted and the single link cluster method is used.

KARLOV

With KARLOV Karhunen–Loeve transformations are performed as a principal component analysis technique. In the 1977 version, the subroutine has been broken down into a variety of routines containing various options. See the manual for these.

Format: KARLOV,NIN,NOUT,NPNT,NPCH $ \$ $ (1976)
Example: KARLOV,11 $ \$ $

The data from NIN equals 11 are used for the analysis and results are printed on NOUT. These are usually then plotted by using subroutine VARVAR.

A typical example from the 1977 version is

KAPRIN,11,12 $ \$ $
KATRAN,12,13 $ \$ $

This example will do the same analysis as the above 1976 one. This should also be followed by VARVAR.

KNN

With KNN, *k* nearest neighbors supervised learning techniques are used to check and predict category memberships.

Format: KNN,NIN $ \$ $ (1976)
KNN,NIN,NPNT $ \$ $ (1977)
NPNT: <0 no training set calculations are listed
$=0$ only data with wrong prediction results and summary statistics are listed for the data set
>0 complete training set output listed
Example: KNN,12 $ \$ $

Distance data must be present on input unit 12. In the 1976 version, all training set and test set results are listed. In the 1977 version, this is equivalent to an NPNT$>$0. If NPNT$=0$, only cases where wrong predictions exist are listed.

MULTI

MULTI is a multicategory supervised learning classification scheme which develops hyperplanes to separate each category from all other categories.

Format: MULTI,NIN,NPCH,NPMAX,NWV $ \$ $ (1976)
MULTI,NIN,NPNT,NPCH,NPMAX,NWV $ \$ $ (1977)
NIN: it is useful to use scaled data (usually from SCALE)

NPMAX: ≤ 0 a maximum of 2000 iterations is allowed

> 0 results are listed after a maximum of NPMAX number of iterations

Example: MULTI,11,0,3000$

Data are read in from unit 11 and a maximum of 3000 iterations for classification is allowed.

NLM

NLM is an unsupervised learning technique used to reduce *n*-dimensional space into two dimensions while preserving interpoint distances.

Format: NLM,NIN,NOUT,NPNT,NPCH,NTRAN,IPOWR,LIMIT,INIT $ (1976)

NLM,NIN,NOUT,NPNT,NTRAN,IPOWR,LIMIT,INIT $ (1977)

NIN: must be distance similarity measurements (usually from *DISTANCE*)

NTRAN: degree of feature space mapped onto (default = 2)

LIMIT: number of iterations allowed (default = 30)

Example: NLM,11$

Distance data are read in from unit 11. A nonlinear mapping of the data from *n*-dimensional space (*n* equals the number of variables considered) to two-dimensional space is made, using up to 30 iterations to minimize the error.

PLANE

PLANE uses a linear discrimination function to separate cases from two distinct categories. A hyperplane is sought which best separates the pair.

Format: PLANE,NIN,NPCH,NPMAX,NWV $ (1976,1977)

NPMAX: number of iterations per category (default = 500)

Example: PLANE,10$

A hyperplane is sought which best separates the pair of categories, using data from input unit 10. The maximum number of iterations allowed is 500. If more than two categories are present in the data, the process and results are shown for all possible pairs of categories.

SCALE

SCALE contains several processing methods for scaling the data to give each variable certain weights in the analysis process.

Format: SCALE,NIN,NOUT,NPNT,NPCH,NSCL$ (1976)
             SCALE,NIN,NOUT,NPCH,NSCL$ (1977)
             NSCL(1976): ≤0 autoscale
                          >0 rangescale
             NSCL(1977): <0 subtraction and division vectors read from cards
                          =0 autoscale
                          =1 mean subtraction
                          =2 variance normalization
                          =3 mean normalization
                          =4 range scale
                          =5 variable weighted autoscale
                          =6 error weighted autoscale
                          =7 error weighted variance normalization
                          =8 error weighted mean normalization

Example: SCALE,10$

Input unit is NIN equals 10 and output is NOUT. The data, after scaling, are not punched, and an autoscaling method, which gives each variable a mean of zero and a standard deviation of one (see Appendix IX) is the method by default.

SELECT

In SELECT features are selected according to their importance in classification.

Format: SELECT,NIN,NOUT,NPNT,NPCH,NWGT,NTRAN $ (1976)
             SELECT,NIN,NOUT,NPNT,NWGT,NTRAN,NCOR,
             TOL$ (1977)
             NWGT(1976): ≤0 correlation to property weights
                          =1 variance weights
                          =2 Fisher weights
             NWGT(1977): same choices as above with the following additional
                          ones:
                          =3 variance weighting by category
                          =4 Fisher weighting by category
             NTRAN: ≤0 feature selection continues until remaining
                        weights are below tolerance
                     >0 NTRAN number of features are selected
             NCOR(1977): ≤0 final features ordered but not orthonormalized
                          >0 final features ordered and orthonormalized

Example: SELECT,11$

Input unit 11 must contain autoscaled data. Correlation to property weights are used to select features. This continues until the remaining ones are below tolerance.

TREE

TREE is an unsupervised learning technique for generating a minimal spanning tree.

Format: TREE,NIN,NPNT,NIT$ (1976,1977)

    NIT: ≤0 the tree will be pruned once with default clustering
      parameters

      >0 the user can define what constitutes a cluster

Example: TREE,11,1$

The data for the tree are read in from unit 11. A full description of the tree is printed, and the default cluster definition is used.

VARVAR

VARVAR produces line printer plots of the data matrix.

Format: VARVAR,NIN,NPRS,NPRO,NPRN,NCAL,IFIND,IFNAM,
    IFCAT$ (1976)

    VARVAR,NIN,NPRS,NPRO,IFIND,IFNAM,IFCAT $ (1977)

    NPRS: <0 no feature–feature plots generated

       =0 all feature–feature plots generated

       >0 user-specified feature–feature plots generated

    NPRO: same as NPRS with feature-property plots

    IFIND: ≤0 no index plots

       >0 index plots

    IFNAM: ≤0 no name plots

       >0 name plots

    IFCAT: ≤0 no category plots

       >0 category plots

Example: VARVAR,10,1,-1,1,1,1 $ (1977)

All feature-vs.-feature plots (and no feature-vs.-property plots) are made for the data on input unit number 10. Three plots are made for each pair: index, name, and category.

WEIGHT

WEIGHT weights the features according to the chosen criteria for importance.

Format: WEIGHT,NIN,NVARI,NFISH,NPROP,NPNT,NPCH,NSEQ$
    (1976)

    WEIGHT,NIN,NVARI,NFISH,NPROP,NPNT,NSEQ$ (1977)

    NVARI: <0 variance weights not calculated

       =0 variance weights stored on default file

       >0 variance weights stored on NVARI file

NFISH: same as NVARI using Fisher weights

NPROP: same as NVARI using correlation-to-property weights

NSEQ: ≤0 no action

>0 transformed data resequenced from highest to lowest weights

Example: WEIGHT,11$

Input unit 11 must contain autoscaled data. The matrix is transformed showing results from all three weighting methods. The new data are stored on the three default files for the process.

III.6. BMDP PROGRAMS

INTRODUCTION

As discussed earlier, each procedure in BMDP is a program itself, and must be accessed individually. See your computer facility for details. To facilitate multiple runs, the control language for each program has been devised so that the same cards with no (or few) changes can be used for most programs. These control cards were discussed in Section III.3 for the air pollution data. In this section, we will assume, as a starting point, those common control cards explained in Section III.3 and illustrated in Table III.3. With the exception of the "/PLOT" paragraph, the control cards preceding the "/END" card will be addressed. This discussion will center on changes necessary for the "/PROBLEM," "/INPUT," and "/VARIABLE" paragraphs from the table. The BMDP programs which found major use in Chapter II will be individually considered with respect to additional instructions within these paragraphs or other additional paragraphs necessary to implement the program. An entry of "no change" means that the paragraph is valid in the form given in Table III.3. A few of the additions to these will be briefly discussed. A more detailed discussion of these and the other BMDP programs available may be found in the manual.

In BMDP, spacing on the computer card is not critical. The following changes may be made utilizing any format chosen. However, paragraph structures should follow those discussed in Section III.3. Note that sentence order is not critical for most paragraphs.

BMDP1D

BMDP1D—univariate statistics.

(a) PROBLEM: no change.

(b) INPUT: no change.

(c) VARIABLE: no change. This choice will calculate the mean, standard deviation, range, standard error of the mean, coefficient of variance, and the largest and smallest raw values, and standard z-scores for each variable.

(d) Options available:
      Other items may be printed by this program. These are defined in a
"/PRINT" paragraph which goes between the VARIABLE and END designations. One choice is to print all the data. This is done by

/PRINT DATA

The data may also be listed with (a) the word "MISSING" replacing any data
piece designated in the "/VARIABLE" paragraph as missing from the original data set; (b) the words "TOO SMALL" and "TOO LARGE" replacing
any data piece smaller or larger than the designations given in the VARIA-
BLE paragraph; (c) a combination of these options. These are implemented
by a "/PRINT" paragraph that lists any combination of these three. The
appropriate words are chosen from

/PRINT MISSING. MINIMUM. MAXIMUM.

The above paragraph will list only those cases having one of the three choices.
To list all of the data and to designate these three choices as well, the paragraph takes on the format

/PRINT DATA. MINIMUM. MAXIMUM. MISSING.

BMDP2D

BMDP2D—detailed data descriptions.

(a) PROBLEM: No change.
(b) INPUT: No change.
(c) VARIABLE: No change. This choice will count and list the distinct values for
      each variable. Univariate statistics for the maximum, minimum, range, vari-
      ance, standard deviation, standard error, mean, median, mode, skewness, and
      kurtosis are listed. A histogram of the data is plotted. A line 130 characters
      wide is given with the minimum value plotted at the far left character and the
      maximum at the far right. In between these are plotted the relative locations
      of the mean, media, mode, quartiles, and standard deviations. Finally, a table
      is listed with an entry for each distinct observed value along with its frequency,
      percentage, and cumulative percentage of occurrence. This must be considered
      critically. If too many distinct values occur in the data, the output here is too
      voluminous and therefor meaningless. This can be alleviated by adding an
      optional "/COUNT" paragraph.
(d) Options available:
      The most commonly used option is a "/COUNT" paragraph placed
between the "/VARIABLE" and "/END" ones. This is used to define how

the frequency table is constructed. Either rounding or truncating the data to certain places can lower the number of distinct values to be listed in the table. For instance, if the fourth through sixth and eighth variables are to be rounded to the nearest 5's, 0.1, 1.0, 10.0 units, respectively, and the ninth variable truncated to the ten's place, the paragraph would read

/COUNT ROUND = (4) 5.0, 0.1, 1.0, (8) 10.0 TRUNC = (9)10.

This is used to change only the listings for distinct values on the table. The statistics printed are then calculated from these truncated values, and may be different from the true statistical values.

BMDP3D

BMDP3D—*t*-test statistics.

(a) PROBLEM: No change.
(b) INPUT: No change.
(c) VARIABLE: A grouping variable must be indicated in the variable list of this paragraph to determine category definitions for the *t*-test analysis. See Section III.3 for details. Each distinct value for this variable indicates a different category. *T*-test analyses are performed for each data variable, utilizing all possible pairs of defined categories. Univariate statistics and histograms are also plotted for each variable.
(d) Options available:

A "/GROUP" paragraph is optional if the grouping variable defined in the "/VARIABLE" paragraph consists of less than ten different group codes (or values). This default assumes that each distinct code indicates a different category. If the grouping variable contains more than ten distinct values, the "/GROUP" paragraph is required to define category structure. See Section III.3 for details.

BMDP4D

BMDP4D—single column frequencies.

(a) PROBLEM: No change.
(b) INPUT: Remove this paragraph's cards.
(c) VARIABLE: Remove these cards. This choice will count the frequency of occurrence of each type of character in each column of the data individually. It will do this only one column at a time. It is therefore utilizing 80A1 format.
(d) Options available

Various printing options for the original data can specified. If a "/PRINT card is placed before the "/END," all of the original data are printed (after the frequency of occurrence counts are printed). A space is inserted after every tenth column in the data. A second printing option can be used to print a subset of the original data. This may involve either the numeric or alphabetic values in the data or the signs, blanks, or symbols. An example would be the inclusion of the paragraph

"/PRINT NUMERIC = ' '.

In this case, all number symbols in the base are replaced by blanks. Only the alphabetic, signs, and symbols (such as commas and periods) are then printed for each pattern. A second example

"/PRINT NUMERIC = 'A'.

would list the data base replacing all number symbols with the letter A. This option is often used for data screening.

BMDP5D

BMDP5D—histograms and univariate plots.

(a) PROBLEM: No change.
(b) INPUT: No change.
(c) VARIABLE: No change.
(d) PLOT: A PLOT paragraph is required. If you want only histograms, the paragraph "/PLOT" is the only requirement. This choice will print one histogram for each variable in the data set. The vertical axis is the histogram base with X's printed along the horizonal axis representing the number of cases in that value interval. Individual and cumulative frequencies and percentage of occurrences are listed for each interval.
(e) Options available:
    Cases may be divided into groups and either plotted in a single histogram with different symbols representing each group, or a number of histograms, each containing some subset of the groups (or each group individually) as specified. For these options, a grouping variable must be designated in the VARIABLE paragraph, and a GROUP paragraph is necessary if the grouping variable takes more than ten distinct values (see Section III.3). A sentence is added to the "/PLOT" paragraph, indicating how the groups are to be considered. Assume that the grouping variable takes on values 1.0, 2.0, 3.0, and 4.0 as with one case when considering the four air pollution sampling stations (MC, MS, NS, and SC). Assume we want to plot MC alone, and then combine MC with NS and SC for a second plot. The "/PLOT" paragraph would read

/PLOT GROUP = 1. GROUPS = 1., 3., 4.

Now, optionally, we may want to do this for only the variables FE and PB. A variable sentence in the PLOT paragraph can be used to define this problem:

/PLOT VARIABLE = FE, PB. GROUP = 1. GROUPS = 1.,3.,4.

The result would be four histogram plots containing FE for MC, PB for MC, FE for MC, NS, and SC combined and PB for the three categories combined. The last two histograms would contain a different symbol on the plot for each of the three categories in order to distinguish them.

The size and scale of the plot may be changed. Also, other types of plots such as normal probability plots, half-normal plots, detrended normal probability plots, cumulative frequency plots, and cumulative histograms may be made. See the manual for these.

BMDP6D

BMDP6D—scatter plots.

(a) PROBLEM: No change.
(b) INPUT: No change.
(c) VARIABLE: No change.
(d) PLOT: A "/PLOT" paragraph is required to specify what is to be plotted and the size of the plots. The program will plot those chosen along with the simple linear regression equations for them. A sentence listing those variables for the y axis and another for the x axis is needed. This takes the form

/PLOT YVAR = FE, MN. XVAR = PB, CO.

Two plots are then made: FE with PB and MN with CO. If the sentence CROSS. is added to the paragraph, all possible pairings (FE with PB, FE with CO, MN with PB, and MN with CO) would be used and plotted. Numbers are plotted as the points with the value representing the number of cases plotted at that same location.

(e) Options available:

An option in the "/PLOT" paragraph is to include the sentence "STATISTICS." This will list the number of points plotted, the correlation between the two variables, the mean and standard deviation for each variable, the regression lines of X on Y and Y on X, and the residual mean square error.

A second option is to change the size of the plot using a "SIZE" sentence. If the sentence "SIZE = 50, 30." were included in the PLOT paragraph, the plot would be 50 characters wide and 30 lines high. The default is "SIZE = 70, 42."

A third option is to specify maximum and minimum values for each variable. These, however, can also be specified in the "/VARIABLE" paragraph.

This option could be used in cases where it is desirable to have each plot generated have identical axis scales. See the manual for this.

A method of grouping cases similar to that in BMDP5D can be used. A grouping variable in the "/VARIABLE" paragraph may be defined or the sometimes optional "/GROUP" paragraph may be used. See Section III.3. A GROUP sentence is then used in the "/PLOT" paragraph to define those to be plotted. Given the grouping variable codes of 1.0, 2.0, 3.0, and 4.0 for MC, MS, NS, and SC, respectively, as in our example, a paragraph written as

/PLOT XVAR = FE. XVAR = MN. GROUP = 1. GROUPS = 1.,2.,4.

would give two plots of FE versus MN, one for MC only and the second for the combination of MC, MS, and SC. A distinct symbol would be used for those cases belonging to each of the three categories in the second example.

BMDP1M

BMDP1M—variable clustering based on similarities between the variables.

(a) PROBLEM: No change.
(b) INPUT: No change.
(c) VARIABLE: No change. This choice will use the absolute value of the correlation as a measure of similarity. Each variable is originally considered a cluster, and these are then combined in subsequent steps according to the minimum distance between any two variables that are not in the same cluster. The mean and standard deviation for each variable is listed along with a summary table of the clusters formed. A tree showing these clusters is printed with an interpretation of the values in the tree.
(d) Options available:
    A "/PROCEDURE" paragraph can optionally be used to select subsets of the original variables to be included in the study. Also in this paragraph the measure of similarity or linkage rule criteria may be altered. The correlation matrix of the variables may also be printed. See the manual for these usages.

BMDP2M

BMDP2M—a cluster analysis for the cases is performed according to some distance measurement.

(a) PROBLEM: No change.
(b) INPUT: No change.
(c) VARIABLE: No change. This choice will use a Euclidean distance to define similarities between cases. The data are first standardized (see Appendix IX) before use. Each case is originally considered a separate cluster. In each step,

the two clusters most similar are combined and considered as a single cluster for the next step. This process continues until all cases become a single cluster. A table summarizing the step-by-step combination of cases is given along with explanation of it. A tree diagram of the clusters is given.

(d) Options available:

A "/PROCEDURE" paragraph can optionally be used to define other measurements of similarity such as chi square or phi square. Weighting of certain cases may also be specified.

A "/PRINT" paragraph optionally allows the standardized data to be printed. The tree may be defined to be printed horizontally instead of the more common vertical output. The initial distances between the cases may also be printed. See the manual for instructions for the use of these optional paragraphs.

BMDP4M

BMDP4M—factor analysis.

(a) PROBLEM: No change.
(b) INPUT: No change.
(c) VARIABLE: No change. If no additional specifications are listed, the program will utilize the default options. The correlation matrix is factored and a principal component method of analysis is used to extract the initial factors. Only those whose eigenvalue is greater than 1.0 are retained. A varimax rotation is performed on the initial factors with a maximum number of rotation iterations being 50. The initial communality values in the correlation matrix are left unaltered. The Kaiser's normalization procedure is performed on the rotated data.
(d) Options available:

Many options exist for this program. Many are specified in a "/FACTOR" paragraph. The method of initial factor extraction can be altered with a "METHOD =" sentence. Choices are MLFA (maximum likelihood factor analysis), LJIFFY (Kaiser's Second Generation Little Jiffy), and PFA (principal factor analysis). If MLFA or PFA is specified, the number of iterations can be defined by an "ITERATE =" sentence (default is 25). The initial communality estimates are changed with a "COMMUN =" sentence. Choices include SMCS (squared multiple correlations) and MAXROW (maximum value in row). The number of factors obtained can be specified by "NUMBER = #." where # is the value for the number to be extracted.

A second optional paragraph is the "/ROTATE" one. A "METHOD =" sentence allows a variety of rotational methods to be chosen, which include VMAX (varimax), DQUART (direct quartimin), QRMAX (quartimax),

EQMAX (equimax), ORTHOG (orthogonal with gamma), DOBLI (direct oblimin with gamma), ORTHOB (orthoblique), and NONE (no rotation). A maximum number of rotations other than the defaulted value of 50 can be assigned with a "MAXIT = #." sentence.

Normal results include data for the first five cases, univariate statistics, the correlation matrix, the squared multiple correlation of each variable with all other variables, the eigenvalues of the factors, the unrotated and rotated factor loadings, plots of the rotated factor loadings, factor score coefficients, variances, and covariances, and estimated factor scores and Mahalonobis distances. Some of these may be eliminated from the printout by using a "/PRINT" paragraph. The raw data for more than the first five cases may also be desired. The standard scores may be requested.

BMDP6M

BMDP6M—canonical correlation analysis.

(a) PROBLEM: No change.
(b) INPUT: No change.
(c) VARIABLE: No change.
(d) CANONICAL: A "/CANONICAL" paragraph is needed to specify the variables to be included in the sets, each of which must contain at least two variables. The two sets are listed, for example, as

/CANONICAL FIRST = PB, CO, FE. SECOND = MN, ORG.

This will then compute and print out the canonical correlations, the coefficients for the canonical variables and the correlations of these to the original variables. The eigenvalues associated with these are given and a test of significance is performed after each is extracted on the remaining eigenvalues. The raw data for the first five cases are also listed. For each original variable, the mean, standard deviation, and smallest and largest observed values are listed.

(e) Options available:
The number of cases to be listed can be increased to more than five by a "/PRINT" paragraph. For example, the paragraph

/PRINT CASE = 20.

indicates that the data are to be printed for the first 20 cases. This paragraph can also be used to print a variety of matrices present in the data calculations as well as t-tests for the regression coefficients and residuals present. See the manual for details.

The "/PLOT" paragraph can be used to plot any variable or residual. The residuals are designated by an R followed by the first seven letters of the name of the dependent variable. Therefore, a designation of

/PLOT XVAR = FE, RFE. YVAR = PB, RPB.

would plot FE versus PB and the residual of FE versus the residual of PB values. The correlation matrix is also given, along with the squared multiple correlations (R^2).

The canonical variables can be plotted to visualize the degree of correlation. To plot the first two canonical variables for each of the two sets of variables against each other a "/PLOT" paragraph is used:

/PLOT XVARS = CNVRF1,CNVRF2. YVARS = CNVRS1,CNVRS2.

CNVRF1 represents the first canonical variable extracted from the first group of original variables. CNVRS1 is the first canonical variable extracted from the second original group of variables. A plot size may also be specified.

BMDP7M

BMDP7M—stepwise discriminant analysis.

(a) PROBLEM: No change.
(b) INPUT: See below.
(c) VARIABLE: A grouping variable must be specified in this paragraph. Also, either a "/GROUP" paragraph must be used to indicate the division of this variable (if there are more than ten distinct values) or the number of groups present must be indicated in the "/INPUT" paragraph with a sentence "GROUP = #." where # is the number of groups to be formed.

This program will list the mean, standard deviation and coefficient of variance for each variable, calculating these for each group individually and then the entire data set. Results after each step in the analysis are listed. After the last step, a summary table of the results is given. Each case is then classified on the basis of these results. Eigenvalues of the matrix are computed and the canonical variable information given. The group means are plotted in a scatter plot followed by a second plot of these plus all cases. The axes are the first two canonical variates.

(d) Options available:
The discriminant analysis may be performed on a subset of groups. A "USE =" sentence in the "/GROUP" paragraph specifies the names or subscripts of the groups used to estimate the classification functions. Groups not

included in this statement are entered into the problem as a test step and classification results are predicted from the training step.

A "/DISC" paragraph can be used to specify an ordering for variable inclusion into the equation. A "LEVEL = " sentence assigns priority levels to the variables. A "METHOD = " sentence is used to determine how variables are to be removed from the function. Limits for the F-to-enter and F-to-remove can be specified. Other stepping methods can be requested in this paragraph.

BMDP1R

BMDP1R—Multiple linear regression.

(a) PROBLEM: No change.
(b) INPUT: No change.
(c) VARIABLE: No change.
(d) REGRESS: A paragraph is needed to define the regression problem. A single dependent variable (or one to be predicted) must be given along with a list of independent variables (or ones used in the prediction equation). This paragraph may be repeated before the "/END" for multiple analyses. For the air pollution data, assuming FE is to be predicted from values of PB, CO, and ORG, the paragraph would take the form

/REGRESS DEPEND = FE. INDEPEND = PB,CO,ORG.

This requests the development of an equation with the form

$$FE = a + b_1(PB) + b_2(CO) + b_3(ORG) + \epsilon$$

where a is the intercept for the equation, ϵ is the error, and b_1, b_2, and b_3 are the coefficients for the variables. If no INDEPEND list is given, all variables besides the DEPEND one and any grouping variables are included. Univariate statistics for each variable (mean, standard deviation, coefficient of variance, and maximum and minimum values) are given. The multiple correlation R, multiple R^2, and standard error of the estimate are calculated. An analysis of variance table with the regression and residual sums of squares and F significance test is listed. A summary table for the regression with the standardized and unstandardized coefficients is also shown.

(e) Options available:

If a grouping variable is defined in the "/VARIABLE" paragraph, or if a "/GROUP" paragraph is used to define category structure, a regression for all groups combined followed by one for each group separately is made. A test is performed to check the quality of these regression lines.

A "/PRINT" paragraph can be used. If one wishes to see the data, residuals, and predicted values, the paragraph would read

/PRINT DATA.

If a sentence "CORR." is included, the simple correlation matrix will also be printed.

A "/PLOT" paragraph can be used for residual plots. This is done with the paragraph

/PLOT RESID.

The size can also be specified.

If a xero intercept is required in a particular problem, a sentence in the "/REGRESS" paragraph can define this as "TYPE = ZERO."

Many other options are available with this program. See the manual for details.

BMDP2R

BMDP2R—Stepwise linear regression.

(a) PROBLEM: No change.
(b) INPUT: No change.
(c) VARIABLE: No change.
(d) REGRESS: A paragraph is needed to define the regression problem. A single dependent variable (the one predicted) must be given along with a list of independent variables (ones used in the prediction equation). This paragraph may be repeated for multiple analyses. If no independent list is given, all variables besides the DEPEND (and any grouping variable) are included. Given the dependent to be FE and the independent to be PB, CO, and ORG, the paragraph would read

/REGRESS DEPEND = FE. INDEPEND = PB, CO, ORG.

The program would utilize a stepwide regression procedure to develop an equation of the form

$$FE = a + b_1(PB) + b_2(CO) + b_3(ORG) + \epsilon$$

where a is the intercept, ϵ the error, and b_1, b_2, and b_3 the variable coefficients. Only those terms above a given criterion for entrance are generated and retained. The mean, standard deviation, skewness, kurtosis, smallest and largest observed values, standard score, and coefficient of variance statistics are listed for each included variable. In this program, the variables are stepped into the equation one by one according to a user-selected criterion (the default is dictated by F-to-enter and F-to-remove limits). Results are printed after each step of the analysis. The multiple correlation R value, multiple R^2, adjusted R^2, and the standard error of the estimate are given. The analysis of variance results and statistics after each step regarding both these variables already included in the equation and those not yet entered into it are listed.

(e) Options available:

Other methods for either entering or removing data may be chosen through a sentence in the "/REGRESS" paragraph. Both forward and backward stepping may be defined. See the manual for details.

The intercept may be defined as zero with a "TYPE = ZERO." sentence in the "/REGRESS" paragraph. The order of variable entry may also be defined in this paragraph with a LEVEL sentence. Each variable is assigned a value related to its level of inclusion. Those with the smallest values are included first in the analysis, and subsequent additions of variables are made according to the order of increasing LEVEL value. For variables assigned the same LEVEL value, inclusion is determined by the method of entering or removing data defined for the problem.

Many other options exist. The number of steps in the analysis may be limited. Various printing choices are available. The data, residuals, and predicted values can be listed by use of a paragraph

/PRINT DATA.

A residual plot may also be specified as

/PLOT RESID.

Other options can be found in the manual.

BMDP4R

BMDP4R—Regression on principal components.

(a) PROBLEM: No change.
(b) INPUT: No change.
(c) VARIABLE: No change.
(d) REGRESS: The "/REGRESS" paragraph is used to describe the dependent variable (the one to be predicted) and the independent variables (those used in the prediction). This paragraph can be repeated for multiple analyses. An example would be

/REGRESS DEPEND = CO. INDEPEND = FE, ORG, PB.

If no independent list is given, all variables except the dependent one are included. A regression analysis is performed for CO on a set of principal components computed from FE, ORG, and PB. The standardized variables are often used. The regression is performed stepwise. The results are given both as principal components and in terms of the standardized variables. For each variable, the mean, standard deviation, and coefficient of variation are calculated. The correlation matrix is listed and the summary of sums of squares given. The F-value, R^2 multiple correlations, and various correlations are listed. The principal component results are summarized.

(e) Options available:

The scores for each case can be listed by

/PRINT SCORE.

Plots can also be made of residuals by

/PLOT RESIDUAL.

which causes plots to be made of the residuals and the residuals squared versus the predicted values. Original variable plots versus predicted values and normal and detrended normal probability plots of the residuals can also be specified.

Criteria for entering variables into the regression equation can be specified by a sentence in the "/REGRESS" paragraph. Choices are CORR. (enters according to the size of the correlation coefficient) or EIGEN. (enters according to the size of the eigenvalue).

BMDP6R

BMDP6R—Partial correlation and multivariate regression.

(a) PROBLEM: No change.
(b) INPUT: No change.
(c) VARIABLE: No change.
(d) REGRESS: A "/REGRESS" paragraph is required to specify the dependent variables (usually more than one) and the independent variables. This takes the form

/REGRESS DEPEND = FE, PB. INDEPEND = CO, ORG.

The partial correlations of FE and PB are calculated removing the linear effects of CO and ORG. The computations necessary include the regression coefficients. Results are printed for the mean, standard deviation, coefficient of variance, smallest and largest observed value and standard score, the skewness, and the kurtosis for each variable. The correlation matrix and the squared multiple correlations are given along with the partial correlations.

(e) Options available:

The program will list the raw data for the first five cases of data. A "/PRINT" paragraph can be used to change this number. You may also select the matrices to be printed in this paragraph. *T*-tests for the regression coefficients are also available.

An optional "/PLOT" paragraph will allow any variable or residual to be plotted against any other variable or residual. Normal probability plots can also be made. The data in each case may also be weighted by a WEIGHT variable if the problem so dictates. See the manual for details.

BMDP3S

BMDP3S—Nonparametric statistics.

(a) PROBLEM: No change.
(b) INPUT: No change.
(c) VARIABLE: No change.
(d) TEST: A "/TEST" paragraph is necessary to specify the statistics to be computed. The choices (and their sentence designation in parentheses following the names) are as follows:
(1) Sign test (SIGN.).
(2) Wilcoxon signed rank test (WILCOXON.).
(3) Friedman two-way analyses of variance and Kendall coefficient of concordance (both accessed by FRIEDMAN.).
(4) Kruskal–Wallis one-way analysis of variance and Mann–Whitney rank sum statistic (the second is used only when the number of groups present is two; both are accessed by KRUSKAL.) If this is designated, no other statistic will be computed (even if requested).
(5) Kendall rank correlation coefficient (KENDALL.).
(6) Spearman rank correlation coefficient (SPEARMAN.). A subset of variables to be used in the analysis can be defined in this "/TEST" paragraph. If no specification is given, all variables are included, and the tests are performed on all possible pairings of the variables. The paragraph takes the form

/TEST SIGN. WILCOXON. KENDALL.
VARIABLES = FE, PB, CO.

This will perform those tests listed under (1), (2), and (5) above on all possible pairs of the variables FE, PB, and CO. Univariate statistics (mean, standard deviation, minimum and maximum values) are given for each variable.

(e) Options available:

A grouping variable may be defined in the "/VARIABLE" paragraph, or a "/GROUP" paragraph may be included. Statistics are then calculated for each specified subset of data from the original matrix.

IV

NATURAL SCIENCE APPLICATIONS

INTRODUCTION

The possible applications of pattern recognition techniques to the natural sciences are limited only by the human imagination. All facets of science are generating reams of data which increases the need for more sophisticated methods of data reduction and analysis. The possibilities are infinite. Therefore, it is not the purpose of this chapter to give an exhaustive list of approaches and applications, but rather to give an idea of the possibilities that exist, and some insight into reference source materials.

Appendix I lists some pertinent books on the subject of pattern recognition and applied statistics. The field, however, develops faster than books (including this one) can be published. The scientist should, therefore, rely heavily on applied journals in his or her field to keep informed. Table IV.1 lists some of the journals which publish a large variety of statistical and pattern recognition articles.

One very excellent literature source is the fundamental reviews published in *Analytical Chemistry* during April of even numbered years. The 1972, 1976, and 1980 issues had exceptionally good reviews on statistical applications. Organization was based on the tool used, not field of application. These reviews are highly recommended (Currie *et al.*, 1972; Shoenfeld and De Voe, 1976; and Kowalski, 1980).

A second literature source to use is the *Computer and Control Abstracts* which devotes an individual section to pattern recognition techniques. Applications to all types of scientific and nonscientific problems are listed. For scientific studies, *Chemical Abstracts* should also be utilized.

BIOLOGICAL APPLICATIONS

Many biological topics have lent themselves to pattern recognition techniques for data analyses. In protein structures, three-dimensional electron density maps have been reduced into sets of connected thin lines which follow the density. These are then defined

TABLE IV.1
Journals Publishing Statistical Related Papers

American Journal of Medicine
Analytica Chimica Acta: Computer Techniques and Optimization
Analytical Chemistry
Artificial Intelligence
Computers and Biomedical Research
Computers in Biology and Medicine
Computers in Chemistry
IEEE Transactions on Computers
IEEE Transactions on Systems, Man, and Cybernetics
Journal of Chemical Information and Computer Science
Mathematical Biosciences
Mathematical Geology
Pattern Recognition

and their patterns recognized. An application to a 2-Å density map of RNAases gave a good view of the enzyme's three-dimensional folding. This skeleton representation can prove to be useful in automated determinations of enzyme shapes (Greer *et al.*, 1974).

The prediction of α-helical regions in proteins using pattern recognition was successfully undertaken. For each protein there are certain functions of the amino acid sequences of the protein segments which permit identification of the protein as either α-helical or non-α-helical. The sequences were coded and pattern recognition techniques were used to separate the two groups (Greer, 1974). This type of study was extended to include secondary structures in the proteins such as turns in the polypeptide chains (Denisov, 1975). Sequence determinations of di-, tri-, and tetranucleotides have been made (Burgard *et al.*, 1977). Induced monomer and dimerization of the monomer–dimer equilibrium in proteins have also been investigated (Horiike *et al.*, 1977).

In botany, discriminant analysis was used to describe relationships between leaf mineral elemental compositions in nine different plant species (Garten, 1978). Three major principal components were extracted from the data on 100 plant species. These included (a) a nucleic acid–protein set related to the phosphorus, nitrogen, copper, sulfur, and iron content; (b) a structural photosynthetic factor correlated to magnesium, calcium, and manganese; and (c) an enzyme factor related to manganese, potassium, and magnesium content.

The chemotaxonomy of Paleozoic vascular plants also was studied (Niklas and Gensel, 1976). Cluster analysis was used. Plant fossils were chemically analyzed and chemotaxonomically related to vascular and nonvascular plant fossils which presently have no established taxonomic affinities. With certain weighting of more important features in the data, the material could be separated into clusters that represented the plant type. It was found that phenolic and monohydroxycarboxy acid constituents indicated vascular plants while variations in carbon chain lengths served to separate vascular and nonvascular plant fossils. Other botany applications include methods to determine growth relationships to plant breeding (Whitehouse, 1969). Leaf contours have been studied. Leaf shapes were reduced to a series of i's and o's with i's being used on boundaries and the interior, while o's were used for exterior points. Each leaf was then digitalized into a grid form, and this grid form was translated into pattern recognition computer parameters to identify leaf types. Venations of leaf interiors were added to increase identification abilities (Billson and Dunn, 1970).

Zoological studies have also seen applications. Afferent visual effects of cats have been studied and pattern recognized (Eckhorn and Poepel, 1974). In humans, the visual space has been broken down into variables that are descriptive of fixed nerve nets. The relationship of these to the visual space has been studied (Leibovic et al., 1971). In entomology, an object classification system was designed to count large samples of field collected insects (Vilela et al., 1976). A computer with color vision capabilities was used to recognize objects based on features describing size, shape, the color spectrum, and the color connectivity of the collected samples. Similar to this, general biological shapes have also been studied (Young et al., 1974). Even honey bees have been pattern recognized (Cruse, 1974).

A pattern recognition study of a beef heart was undertaken. The goal was to use chemical data to determine the anatomical site from which the tissue was taken. The training set included emission spectroscopic measurements for Pb, Cu, Mo, Sr, Ce, Ba, and Al for tissue samples taken from 15 anatomic sites in the beef heart. Eight-five per cent of the samples were correctly classified by discriminant analysis with respect to their anatomic origin (Webb et al., 1976b). A later similar study was made on dog heart tissue differentiation (Webb et al., 1976c).

Biological time series have received much attention in the pattern recognition field (Wong et al., 1976). Statistical criteria for the measure of "significance" to study the periodicities as well as the irregular high-frequency oscillations in the time series were made. Hormonal time series studies indicated a real usefulness in this type of analysis for complex multivariate biological systems. A review (with references) was published describing the use of multivariate Gaussian statistics in biological systems (Winkel and Statland, 1975). The problems with interindividual variabilities and stabilities over time was discussed. The problem centers on the changeability of biological samples so that measurements made on these for classification purposes reflect this uncertainty. This makes pattern recognition techniques more difficult to apply, since data consistent over time are not present. Some type of "uncertainty" must then be reflected in the features used.

Many studies have been performed on classifying chromosomes. In one paper, algorithms were developed for classifying these images as one of three types of chromosomes based on shape-oriented features. Correct classifications using the computer exceeded most manual grouping (Lee, 1975).

A HEMATRAK system was discussed as a pattern recognition device for classifying and counting while blood cells (Tebbe, 1976). A minicomputer was connected to a microscope, flying spot scanner, and a search and capture mechanism. This paper describes the major design strategies in the software of the computer to optimize the statistical approach. A second paper describes a real-time high-speed system (Neurath *et al.*, 1973). The discriminating power of the algorithm has been tested to be between 67% and 92% (depending on the definition of accuracy) to identify the number of white blood cells present.

Other applications in biology include a successful computer simulation of nerve cell functions (Tsukada *et al.*, 1975), tests of liver functions (Salkie and Luetchford, 1976), fungi population analyses (Kent, 1972), and applications to phase problems of X-ray diffraction studies of biological membranes (Luzzati *et al.*, 1972). Papers discussing general applications in microbiology (Gyllenberg and Eklund, 1967) and biogenesis and evolution (Yockey, 1973) were found.

MEDICAL APPLICATIONS

Medical fields have seen a large surge in the applications of pattern recognition. Normally, a doctor observes a syndrome and compares it with the identification of syndromes found in his or her knowledge and experience. The diagnosis is then given as the syndrome which is most similar to the one in question. A "nearest-neighbors" pattern recognition technique with the computer was used to mimic this type of doctor's procedure. The program was applied so that when it was used as this type of diagnostic machine, the criteria for prediction could be altered continuously as the result of learning (Ben-Bassat, 1976). In a few papers pattern recognition was used to classify patients' characteristics from questionnaires (Williams, 1976; Batchelor, 1976). One problem undertaken was to combine data from the patient's state of health and clinician's state of knowledge (Gheorghe *et al.*, 1976). Feature selection for treatment decisions were made. The cost functional for the decision process was optimized.

Small computers have recently been used to reduce workloads for medical professionals and to improve the quality of health care by the use of minicomputer-based pattern recognition. An example is the use of chest thermograms to detect early signs of breast cancer (Dwyer *et al.*, 1976). The thermogram as converted into a grid system which as coded and fed into computers. Studies were made to compare the sections of the grid system for normal and abnormal cases. One hundred prospective features were chosen for distinguishing the two cases. Through feature selection techniques, 13 were finally chosen.

Other similar diagnostic uses studied chest photofluorograms, especially abnormal shadows, to diagnose lung cancer and lung tuberculosis (Suenaga *et al.*, 1974). Optical digital pattern recognition on 64 chest films of normal and abnormal lungs were tested (Stark and Lee, 1976). Five feature selection algorithms were applied to the data in order to determine their ability to diagnose coal workers' pneumoconiosis. Machine recognition has also been used in pathology. Tissue sections of tumor areas were used to distinguish malignant, normal, and benign samples automatically (Mancillas and Ward, 1975). Cervical smear characteristics were studied for the development of an automatic prescreening instrument for diagnosing cervical cancer (Imasato *et al.*, 1975). Normal and abnormal smears were compared.

A major medical application for pattern recognition techniques occurs in the field of pharmacology. Drug types have been separated (Sotzberg and Wilkins, 1976). In one study, X-ray diffraction data were used to distinguish 89 compounds into their medicinal use as either sedatives or tranquilizers. Mass spectral data of these same two types were also investigated (Ting *et al.*, 1975). Nonlinear mapping and nearest-neighbors pattern recognition techniques were used to distinguish these two. Pharmacological activities of other test drugs were then predicted with high accuracies. In another study a binary pattern classifier was used to categorize tranquilizers and sedatives. Two hundred and nineteen drugs were used with 69 descriptors for each (Stuper and Jurs, 1975). These descriptors included three types: (a) numeric and binary fragment descriptors. (b) binary substructure descriptors, and (c) topological descriptors. The two-dimensional structure diagram of the drug molecule was used for input data into the system. The two drug classes were linearly separable and all were correctly classified. Feature selection was then performed to extract the 30 best descriptors. One hundred per cent prediction was still possible. Other drugs whose classifications were known were then fed into the system as "unknown" test patterns. Prediction success was 90%. Factor analysis was also used to derive structure–activity relationships. Compounds included were antihistamines, anticholinergics, analgesics, antidepressants, antipsychotics, and anti-Pakinsonism agents.

In another study 200 drugs previously tested by the National Cancer Institute for activity in the solid tumor adenocarcinoma 755 screening tests were studied (Kowalski and Bender, 1974). A 93.5% accurate prediction was obtained in separating those drugs with positive antineoplastic activity from those with no anticancer effects. A more rational approach to drug design was suggested.

Antineoplastic activity of 24 test compounds in experimental mouse brain tumor systems was predicted (Chu *et al.*, 1975). The training set used consisted of 138 structurally diverse compounds tested in these tumor systems. The molecules were represented by their substructural fragments. A nearest-neighbors approach showed an 83% correct predictability while learning machines were right 92% of the time.

Three-dimensional structures of molecules were related to biological activities by the use of principal component analyses (Dierdorf and Kowalski, 1974). The three-dimensional coordinates were coded into computer compatible number and auxilary data such as electronegativities for each atom in the molecule were added. Pattern recognition techniques were then used to relate the data to carcinogenicity of the molecules.

Eight naturally occurring mixtures of estrogenic steroids were analyzed (Smith *et al.*, 1973). A computer interpretation of the mass spectral data obtained was used to study structure specificity.

In other medicinal applications, fetal heart rate patterns were studied using linear discriminant functions and nonparametric algorithms (Klinger and McDonald, 1975). Seventeen samples of five vectors each were used as the training set. Coordinates for each sample vector were features obtained from the fetal heart rate curves and the uterine contractions pressure data.

Baeysian statistics and linear discrimination functions were used to study patterns in plasma renin activity and plasma aldosterone concentrations in 60 patients. Measurements were taken in recumbancy, and before and after intravenous administration of diuretics under various conditions. Important features become those that changed with this cause/effect model (Lui *et al.*, 1976).

Multivariate analyses were used to analyze emission spectroscopic data for Cu, Fe, Al, Ni, Sr, Ba, Mn, Cs, Sn, Cr, Zn, Pb, Mo, and Cd in blood serum of patients showing signs of acute myocardial infarction and other cardiac trauma (Webb *et al.*, 1976a). Computer analyses showed that Cu, Fe, Al, and Ni changed significantly during the first seven days following infarction and could be used to classify patients. A discriminant analysis study was successful to 91%. This was raised to 100% if age, race, and sex were included pairwise with the four other elements.

Using nuclear medicine, direct measurements can now be made of body processes in humans that could previously be examined only in experimental animals. This creates large amounts of data such as the distribution of the quantities of radiation in specific sites in the body over periods of time. These data must then be reduced and analyzed to extract meaningful results from it. A paper discussing the use of statistical techniques and data analyses by computer addresses this problem (Wagner and Natarajan, 1972).

An overall review of the use of computer techniques in medicine discusses many of the problems of such applications (Rutovitz, 1972). Computers were once used only to recognize and count different types of cell organisms. Now more usage is being made of pattern recognition to classify and separate objects into a discrete number of groups. Coding the data from medicinal characteristics to computer compatible numbers is the most difficult step in the process. This requires much skill and practice and it is through success at this step that meaningful results can be obtained. This paper addresses many problems and suggestions for this step.

Other medicinal applications include the use of pattern recognition in studying human hormones (Liu *et al.*, 1975), dental X- rays (Levine *et al.*, 1970), asthma (Bordas *et al.*, 1975), and surgery decisions (Patrick *et al.*, 1975).

GEOLOGY AND EARTH SCIENCE APPLICATIONS

Geological and earth sciences have also found the applications of computer techniques very advantageous for data reduction problems. Pattern recognition has been used for uranium prospecting. A quantified decision process selecting patterns of geological data that are known uranium sites can be used for selecting other areas where yet undiscovered deposits of uranium (or other metals) are likely to be found (Briggs and Press, 1977; Ackermann, 1973). Copper deposits were also studied (Olade and Fletcher, 1976). Many types of data from standard geological maps, geophysical, geochemical, and radiometric surveys, as well as satellite imagery are now used in commercial explorations. These can be made computer compatible and pattern recognition techniques applied for prediction processes. An exploration paper describes the results of a computer simulation experiment to relate frequency distributions and spatial distributions of elements. Questions related to physical size and sampling patterns are raised. Developments in computer data presentations and interpretations, including pattern recognition sorting and classifications are discussed (Govett *et al.*, 1975).

Pattern recognition has also been used to analyze standard geological data to identify areas in California and Nevada that are particularly earthquake prone (Briggs *et al.*, 1977; Gelfand and Guberman, 1976). An algorithm is used to define suites of characteristic traits and can successfully distinguish earthquake characteristics of the San Andreas fault system from those characteristics of the western Basin and Range province. It also pinpoints areas unlikely to be the epicenter of strong earthquakes. Predictions of future earthquake epicenters are made. Features characteristic of earthquake-prone areas emerge that appear physically meaningful in terms of large-scale geology. Other papers on earthquake recognitions have also been published (Picard and Sallantin, 1977; Puzyrev and Gol'din, 1975).

Many papers have been written on the classification of geological materials using pattern recognition techniques. In one, large sets of rock compositions were grouped in multi- dimensional space according to chemical variables (Brotzen, 1975). The point density around each composition was estimated. Establishment of the identity of specimens contained in these clusters and systematic comparisons of the point densities surrounding each of them leads to the recognition of density maxima, which represent the statistical modes of the rock types present in the set. The remaining specimens can then be assigned to the group to which they are most similar.

Samples from Lake Geneva were analyzed for 29 oxides and trace elements. Cluster analyses, data transformations, and data reductions were used and their results compared for predictive ability. As a final result, the 29 original variables were reduced to four components and the sediment samples classified into four facies, leading to easily interpretable geochemical maps (Jaquet *et al.*, 1975).

Regression analyses were applied to trace element data from stream sediments in New Brunswick. Regression-predicted values of Zn, Pb, Cu, and Mo were compared to observed values. Chemical anomalies in the results were studied and related to geological relief maps. Factor analysis was applied to 158 stream sediment samples in Canada and geological significance studied (Saager and Sinclair, 1974). A Pb–Zn–Ag factor correlated with volcanic rock and certain veins known to be present in those rocks. The second factor, high in Cu and Mo, correlated with porphyric intrusions. The third, Sb–Ag, could not readily be explained in geological terms.

Granite rock intrusions were mapped using cluster and discriminant analyses (Rhodes, 1969). These methods were compared to the normal largely subjective approach of conventional field mapping. Two examples were given. In one, there is good agreement between the numerical and field methods. In the second, the numerical approach shows two chemically distinct granites in what was believed, on the basis of field and petrographic evidence, to be a single homogeneous granite.

Recently, pattern recognition applications to lunar classification studies have been made. Lunar science, a field barely ten years old, has produced a vast amount of major and minor elemental data on the samples returned during the Apollo missions. Papers characterizing rock types present were published (Pratt *et al.*, 1977; Pratt *et al.*, 1978). Correlations between the different areas of the moon were also made.

With the lunar data, no preconceived ideas of the number of groups expected were formed so group definition had to rely totally on unsupervised learning. Rock types similar to the ones found on the earth were shown using both major and elemental data (a total of up to 23 features). Subcategories within rock group types were also found. After these were defined, other lunar rock analyses could be added and the sample assigned to a group. Relationships between the different landing sites were also investigated.

A similar study was made on characterization and categorization of stony iron and iron metorites. Pattern recognition techniques were used on trace elemental abundance data to define meteoritic groups (Pratt *et al.,* 1977). Petrological features were added. Preaccepted category definitions existed for this data base, so supervised pattern recognition was used to check these previous definitions and to assign categories to newly analyzed meteorites. It was also possible to select the features that contributed most to the category separations.

Applications in geology to mineralizations have been made. A recent study on gold distributions in South African ores was completed (Krige, 1976). Trace elemental analyses in chrome-ore bodies have also been utilized to characterize the mines (Varma and Jaipuriar, 1977). Drilling resistance classifications of rocks have been made (Komarov *et al.,* 1975). Soil correlation studies were made to relate the Mn content with soil properties (El-Leboudi *et al.,* 1971).

Volcanic studies were another field of application. Petrochemical and petrogenetic studies of volcanic associations were made (Karche and Mahe, 1977). Recent volcanoes in the Central Andes were also studied (Frangipane-Gysel, 1977).

Geostatistics is becoming a rapidly growing field. Statistical modeling in geology has increased (Wiatr *et al.,* 1974). The normal versus log-normal distributions have been compared (Lucero, 1973). Geostatistics in the atomic energy field have been used (Davids, 1971). Also, alternative experimental procedures for simultaneous multielemental trace analyses for geochemical materials obtained by dc arc emission spectroscopy have been studied (Maessen *et al.,* 1975). Applications have been made to classify engineering and geological environments as well (Wiatr and Stenzel, 1974). The misuse of various regression techniques applied to geochemical problems has also been discussed (Mark and Church, 1977).

ENVIRONMENTAL APPLICATIONS

Environmental problems utilizing pattern recognition techniques generally fit into three topics: air or water pollution and oil spill analyses. Air pollution and source identification studies have been made. Statistical distributions of pollutants in the various areas have been investigated. A study, similar to the one used as an example in this book, was performed on the Southwestern atmosphere near Tucson. Pattern recognition techniques were used to determine pollutant characteristics and sources (Gaarenstroon *et al.,* 1977).

A statistical study of instrumental neutron activation analysis data over a five-month period in the Boston metropolitan area was also made. Factor analysis and hierarchical clustering techniques were utilized (Hopke et al., 1976). Six common factors were found which accounted for 77.5% of the variance in the data.

Multiple regression analyses have been used on x-ray fluorescence data of 15 elemental species in ceramic supported auto emission catalysts (Mencik et al., 1976). An equation was developed to relate the x-ray intensities of the elements and contaminants.

Granulometric distributions of atmospheric aerosols have been plotted. Patterns and trends in these were studied to summarize some of the characteristics that control such distributions (Perone et al., 1976; Renoux et al., 1975).

Atmospheric corrosion of metals was investigated (Mikhailovskii et al., 1976). Causes of the corrosion in relationship to various atmospheric pollutants were identified.

In water pollution studies, waste-water treatment methods were analyzed. Pattern recognition techniques were applied to relate biological, physical, and chemical water qualities (Hamilton et al., 1976; Vandeginste and Van Lersel, 1978).

With the energy crises present today, oil and resulting oil spills have become a problem of concern. One of the major roadblocks to oil spill identifications is the weathering effect seen on the spill samples. Changes in chemical characteristics result. Pattern recognition techniques have been used to sort out such factors so spills can be identified. A variety of chemical spectroscopic techniques have been used, such as infrared, fluorescence, mass spectra, and neutron activation analyses. A conference was held in 1976 to discuss many of the implications from these studies (Workshop on Pattern Recognition Applied to Oil Identification; Coronado, California, November 11–12, 1976, sponsored by the University of Connecticut, U.S. Coast Guard, and IEEE Computer Society).

In another oil spill source identification study, weathering factors were taken into consideration. Field samples were considered as single points in an n-dimensional space utilizing chemical data for the axes. Fuzzy points, or areas of the plot in contrast to single points, were used to define weathered samples. These were then compared for identifications (Duewer et al., 1975).

Multicomponent oils including crude, lubricant, distillates, and residual fuels were analyzed by infrared techniques. The spectra were computer coded and used for binary decisions relating the oil types (Mattson, 1976).

An analysis of oils in natural waters was also made (Lysyj and Newton, 1972). Petroleum pollutants in natural waters were differentiated from organic compounds derived from biomass activity by thermal fragmentations of organic molecules.

PHYSICS APPLICATIONS

In physics, few papers as such using pattern recognition as a whole were found, but isolated techniques included in this book are widely used in the field. A study relating pattern recognition to crystal symmetry in molecules was made (Siromoney et al.,

1973). Two models were discussed—a matrix model which was generalized to n- dimensions and a two-dimensional array model. Pattern recognition techniques were related to symmetry operations. A second application to protein crystalography was also made (Feigenbaum *et al.,* 1977).

Pattern recognition methods were also applied to bubble chamber data (Videau, 1973). The approach was designed to resolve problems concerning processing information by automatic measurements. Filtering methods were considered and feature analysis used to reduce the number of parameters necessary for a pattern.

Nonparametric methods were used on γ-ray astonomical experimental data to identify events. The training set included labeled events and classifications were made by introducing pre-established linear threshhold decision functions (Buccheri *et al.,* 1975).

Pattern recognition was also used to classify curium energy levels. Features utilized in the training set were the energy levels, Lande g values, J values, and isotope shifts. Forty levels were identified with high probabilities and 14 more were assigned to one or two possible configurations (Peterson *et al.,* 1978).

Studies have also been made on Al-rich alloys to relate the chemical compositions to the time of failure (Ganguly and Dhindaw, 1977). Accelerated stress corrosion cracking tests were performed on alloys containing variable amounts of zinc (4–6%), magnesium (0.5–1.5%), and copper (0.5–1.5%). Regression analyses were used. It was found that the time to failure increased with increasing copper content and decreasing magnesium and zinc.

CHEMICAL APPLICATIONS

In chemistry, the applications of pattern recognition are unlimited. This field of science realized early the need for data interpretation techniques as the amounts of data created grew. The literature and applications in this field are enormous. Reviews on the applications of pattern recognition to chemistry are numerous (Kowalski and Bender, 1972b; Kowalski and Bender, 1973; Kowalski, 1974; Jurs and Isenhour, 1975; Pratt *et al.,* 1978). A recent book was also published from the papers on pattern recognition at a conference (Kowalski, 1977). A view on chemometrics, the study of chemistry and statistics, was made by Kowalski (1975).

Many papers have concentrated on the use of pattern recognition for molecular structure descriptions. In at least 30 papers, mass spectra have been coded into computer parameters. Descriptor programs have been written to identify fragments, substructures, and geometrical parameters from these (Zander and Jurs, 1975). This problem was begun as early as 1971 (Buchanan *et al.,* 1971). Areas of the spectrum were coded for the presence or absence of peaks. Utilizing spectra of known compounds, the computer was trained to relate certain chemical characteristics of the compound to the presence or absence of peaks. Low-resolution mass spectra of alkanes and alkenes were used (Sasaki and Ishida, 1972). The coding and compounds included in the data set were expanded and more sophisticated algorithms used to train the computer (Bender *et al.*, 1973; Schechter and Jurs, 1973). Applications were made to specific classes of compounds such as steroids (Smith *et al.*, 1972), sedatives and tranquilizers (Ting *et al.*, 1975), and other drugs (Abe *et al.*, 1976). Simplex pattern recognition was also investigated with mass spectra (Lam *et al.*, 1976).

Similar studies utilizing infrared data were made (Liddell and Jurs, 1973; Liddell and Jurs, 1974). In one study binary data were used to indicate the presence or absence of a peak in certain segments of the spectra. Each of these was used as a binary feature and pattern recognition techniques were used to characterize these spectra (Lowry *et al.*, 1975). The test set included 200 spectra for 13 functional groups.

Nuclear magnetic resonance spectra were used similarly. A k-nearest-neighbor classification rule was utilized in one study to relate chemical functionality to spectra (Kowalski and Bender, 1972a). Carbon-13 nuclear magnetic resonance data were used to train the computer to make decisions on the presence or absence of seven structural features. Coded spectra for 62 unknown compounds were then used and results for their features predicted (Wilkins and Isenhour, 1975). Simplex pattern recognition techniques were applied to train the computer for several hundred spectra to predict the presence or absence of three features in the molecule (Brunner *et al.*, 1976). This was then used to interpret 2000 other spectra.

Organic synthetic pathways were studied with pattern recognition techniques (Blower and Whitlock, 1976). The knowledge of successful syntheses was computer coded and trained, and predictions for other syntheses made from these. Iron clathro-chelates metal carbonyl bondings were studied with pattern recognition and the biological activity was related to the molecular structure (Dierdorf, 1974). Known compounds were computer coded and trained. Unknowns were predicted from this knowledge.

In electrochemistry, factor analysis techniques have been applied to polarographic data. Half-wave potentials were used to characterize and pattern-recognize the spectra (Howery, 1972). Overlapped voltametric data were studied (Thomas and Perone, 1977). Simple visual studies of the spectra could not have determined the species present due to the overlap in the peaks. The aid of the computer was necessary to train and test the algorithms relating the spectra and the components present.

Pattern recognition techniques have also been used for data interpretation from various gas chromatography studies. In one study, pyrolysis chromatogram characteristics were coded and studied (Kullik *et al.*, 1976). A linear learning machine was used to recognize different fiber characteristics. Samples used included synthetic, acrylic, polyesters, polyamide, and polyurethane fibers. Mixed retention mechanisms were also studied by factor analyses and linear regression methods (Weiner *et al.*, 1974). Petroleum samples were also pattern-recognized from their gas chromatograms (Clark and Jurs, 1975). Unknowns were then classified according to the samples in this to which they were most similar.

SUMMARY

It is realized that this chapter and book are out of date before publication, owing to the rapid expansions in pattern recognition techniques and applications. We have not attempted to make an exhaustive review, but rather, to give the reader some insight into possible applications. Pattern recognition, correctly applied, can become an invaluable tool in data analysis. It is by no means meant to replace the scientist in the critical data analysis step; but only to aid him or her in making some powerful interpretations already inherently present in the data.

REFERENCES

Abe, H., Kumazawa, S., Taji, T., and Sasaki, S. (1976). *Biomed. Mass Spectrom.* **3**, 151–154.

Ackermann, H. (1973). *Erzmetall* **26**, 161–167.

Batchelor, B. G. (1976). *Third Int. Joint Conf. on Pattern Recogn.* Coronado, California, IEEE, pp. 556–563.

Ben-Bassat, M. (1976). *1976 Joint Workshop on Pattern Recogn. and Art. Intellig.* Hyannis, Massachusetts, 5 pp.

Bender, C. F., Shepherd, H. D., and Kowalski, B. R. (1973). *Anal. Chem.* **45**, 617–618.

Billson, M., and Dunn, R. A. (1970). *IEEE Conf. Record of the Symp. on Feature Ext. and Selection in Pattern Recogn.* Argonne, Illinois, 68 pp.

Blower, P. E., Jr., and Whitlock, H. W., Jr. (1976). *J. Am. Chemical Soc.* **98**, 1499–1510.

Bordas, J., Chomy, P., Childer, H. R., Deboucaud, M., Freour, P., Tessier, J. F., Mallet, J. R., Bernadou, M., and Gachie, J. P. (1975). *Rev. Fr. Allergol.* **15**(3), 175.

Briggs, P. L., and Press, F. (1977). *Nature* **268**, 125–127.

Briggs, P., Press, F., and Guberman, SH. A. (1977). *Geol. Soc. Amer. Bull.* **88**, 161–173.

Brotzen, O. (1975). *Math. Geology* **7**, 191–214.

Brunner, T. R., Wilkins, C. L., Lam, T. F., Soltzberg, L. J., and Kaberline, S. L. (1976). *Anal. Chem.* **48**, 1146–1150.

Buccheri, R., DiGesu, V., and Salemi, S. (1975). *Nucl. Instrum. Methods* **123**(3), 563–572.

Buchanan, B. G., Duffield, A. M., and Robertson, A. V. (1971). *Mass Spectrom.: Tech. Appl.*, 121–178.

Burgard, D. R., Peroue, S. P., and Wiebers, J. L. (1977). *Biochemistry* **16**(6), 1051–1057.

Chu, K. C., Feldmann, R. J., Shapiro, M. B., Hazard, G. F., Jr., and Geran, R. I. (1975). *J. Med. Chem.* **18**, 539–545.

Clark, H. A., and Jurs, P. C. (1975). *Anal. Chem.* **47**, 374–378.

Cruse, H. (1974). *Kybernetik* **15**(2), 73–84.

Currie, L. A., Filliben, J. J., and Devoe, J. R. (1972). *Anal. Chem.* **44**, 497R–512R.

Davids, N. C. (1971). *Librart* **71**, 1106.

Denisov, D. A. (1975). *J. Theor. Biol.* **55**, 107–114.

Dierdorf, D. S. (1974). Dissertation from Univ. Wash., Seattle, Wash., *Diss. Abstr. Int. B* **36**(6), 2793.

Dierdorf, D. S., and Kowalski, B. R. (1974). *U.S.N.T.I.S. AD Rep., Govt. Rep. Announce.* **74**(24), 33.

Duewer, D. L., Kowalski, B. R., and Schatzki, T. F. (1975). *Anal. Chem.* **47**, 1573–1583.

Dwyer, S. J., McLaren, R. W., and Harlow, C. A. (1976). *Joint Workshop on Pattern Recognition and Artificial Intelligence,* Hyannis, Massachusetts, IEEE, 11 pp.

Eckhorn, R., and Poepel, B. (1974). *Kybernetik* **16**(4), 191–200.

El-Leboudi, A. E., El-Sherif, S., and Ismail, A. (1971). *U.A.R.J. Soil Scie.* **11**(1), 77–87.

Feigenbaum, E. A., Engelmore, R. S., and Johnson, C. K. (1977). *Acta Crystallogr., Sect. A.* **A33**, 13–18.

Frangipane-Gysel, M. (1977). *Schweiz Mineral Petrogr. Mitt.* **57**, 115–134.

Gaarenstroom, P. D., Perone, S. P., and Moyers, J. L. (1977). *Environ. Sci. Technol.* **11**(8), 795–800.

Ganguly, R. I., and Dhindaw, B. K. (1977). *Br. Corros. J.* **12**(4), 239–240.

Garten, C. T., Jr. (1978). *Am. Nat.* **112**(985), 533–544.

Gelfand, I. M., and Guberman, S. A. (1976). *Phys. Earth and Planet. Inter.* **11**, 227–283.

Gheorghe, A. V., Ball, H. N., Hils, W. J., and Carson, E. R. (1976). *Int. J. Bio-Med. Comput., Great Britain,* **7**(2), 81–92.

Govett, G. J. S., Goodfellow, W. D., Chapman, R. P., and Chork, C. Y. (1975). *Math. Geology* **7**, 415–446.

Greer, J. (1974). *J. Mol. Biol.* **82**, 279–301.

Gyllenberg, H. G., and Eklund, E. (1967). *Ann. Acad. Sci. Fenn. Ser. A IV Biol.* **113**, 3–16.

Hamilton, T. M., Dalrymple, J. F., Coackley, P., Woodhead, T., and Crowther, J. M. (1976). *Systems and Models in Air and Water Pollution.* 10 pp.

Hopke, P. K., Gladney, E. S., Gordon, G. E., Zoller, W. H., and Jones, A. G. (1976). *Atmos. Environ.* **10**(11), 1015–1025.

Horiike, K., Shiga, K., Isomoto, A., and Yamano, T. (1977). *J. Biochem.* **81**(1), 179–186.

Howery, D. G. (1972). *Bull. Chem. Soc. Jpn.* **45**, 2643–2644.

Imasato, Y., Yoneyama, T., Watanabe, S., and Genchi, H. (1975). *Comp. Biol. Med.* **5**(3), 245–255.

Jurs, P. C., and Isenhour, T. L. (1975). *Applications of Pattern Recognition,* Wiley–Interscience, New York, 184 pp.

Karche, J. P., and Mahe, J. (1977). *Rev. Geogr. Phys. Geol. Dyn.* **19**(2), 125–136.

Kent, J. W. (1972). *Trans. Br. Mycol. Soc.* **58**(2), 253–268.

Klinger, A., and McDonald, I. S. (1975). *IEEE Trans. Biomed. Eng.* **BME-22**(1), 18–24.

Komarov, M. A., Kacherzhuk, S. S., and Rvachev, V. M. (1975). *Izv. Vyssh. Uchebn. Zaved. Geol. Razved.* **5**, 151–155.

Kowalski, B. R. (1974). *Comput. Chem. Biochem. Res.* **2**, 1–76.

Kowalski, B. R. (1975). *J. Chem. Inf. Comput. Sci.* **15**, 201–203.

Kowalski, B. R. (editor) (1977). *Chemometrics: Theory and Application,* ACS-Symposium Series 52, American Chemical Society, Washington, D.C.

Kowalski, B. R. (1980). *Anal. Chem.* **52**, 112R–122R.

Kowalski, B. R., and Bender, C. F. (1972a). *Anal. Chem.* **44**, 1405–1411.

Kowalski, B. R., and Bender, C. F. (1972b). *J. Amer. Chem. Soc.* **94**, 5632–5639.

Kowalski, B. R., and Bender, C. F. (1973). *J. Amer. Chem. Soc.* **95**, 686–693.

Kowalski, B. R., and Bender, C. F. (1974). *J. Amer. Chem. Soc.* **96**, 916–918.

Krige, D. G. (1976). *J. S. Afr. Inst. Min. Metall.* **76**(9), 383–391.

Kullik, E., Kalurand, M., and Keel, M. (1976). *J. Chromatogr.* **76**, 249–256.

Lam, T. F., Wilkins, C. L., Brunner, T. R., Soltzberg, L. J., and Kaberline, S. L. (1976). *Anal. Chem.* **48**, 1768–1774.

Lee, E. T. (1975). *IEEE Trans. Syst., Man, and Cybern.* **SMC-5**(6), 621–624.

Leibovic, K. N., Balslev, E., and Mathieson, T. A. (1971). *Kybernetics* **8**(1), 14–23.

Levine, D. A., Hope, H. H., and Shakun, M. L. (1970). *AFIPS Conf. Proc. 1970 Spring Joint Conf.,* Atlantic City, New Jersey, 487–491.

Liddell, R. W., III, and Jurs, P. C. (1973). *Appl. Spectrosc.* **27**, 371–376.

Liddell, R. W. III, and Jurs, P. C. (1974). *Anal. Chem.* **46**, 2126–2130.

Liu, T. S., Vagnucci, A. H., and Wong, A. K. C. (1975). *Proc. of 1975 Intern. Conf. on Cyb. and Society,* pp. 192–195.

Liu, T. S., Vagnucci, A. H., and Wong, A. K. C. (1976). *Proc. of the IEEE Int. Conf. on Cyb. and Society,* Washington, D.C., 384–388.

Lowry, S. R., Woodruff, W. B., Ritter, G. L., and Isenhour, T. L. (1975). *Anal. Chem.* **47**, 1126–1128.

Lucero, H. N. (1973). *Bol. Assoc. Geol. Cordoba* **2**, 1–13.

Luzzati, V., Tardieu, A., and Taupin, D. (1972). *J. Mol. Biol.* **64**, 269–286.

Lysyj, I., and Newton, P. R. (1972). *Anal. Chem.* **44**, 2385–2387.

Maessen, F. J. M. J., Elgersma, J. W., and Boumans, P. W. J. M. (1975). *Colloq. Spectrosc. Int. (Proc.), 18th* **2**, 530–536.

Mancillas, R. G., and Ward, A. (1975). *Comput. Biol. Med.* **5**, 39–48.

Mark, D. M., and Church, M. (1977). *Math. Geology* **9**, 63–75.

Mattson, J. S. (1976). *U.S. NTIS, AD Rep.:* AD-A039387, 1–48.

Mencik, Z., Berneburg, P. L. and Short, M. A. (1975). *Adv. X-Ray Anal.* **18**, 396–405.

Mikhailovskii, Y. N., Sergeeva, E. I., San'ko, V. A., and Agafonov, V. V. (1976). *Zashch. Met.* **12**, 105–108.

Moore, C. B., Pratt, D. D., and Parsons, M. L. (1977). *Meteoritics* **12**, 314–318.

Neurath, P. W., Brenner, J. F., Selles, W. D., Gelsema, E. S., Powell, B. W., Gallus, G., and Vastola, E. (1973). *Intl. Computing Symposium 1973*, Davos, Switzerland, pp. 399–405.

Niklas, K. J., and Gensel, P. G. (1976). *Brittonia* **28**, 353–378.

Olade, M., and Fletcher, K. (1976). *25 Int. Geol. Congr. (Sydney)* Conf. Dates: 16 August to 26 August, No. 76-0161, p. 635.

Patrick, E. A., Stelmack, F., and Panda, D. P. (1975). *Comput. Biol. Med.* **4**, 293–300.

Perone, S. P., Gaarenotroom, P. D., Moyers, J. L. (1976). *Proc.-Inst. Environ. Sci.* **22**, 477–479.

Peterson, K. L., Anderson, D. L., and Parsons, M. L. (1978). *Phys. Rev. A* **17**(1), 270–276.

Picard, C. F., and Sallatin, J. (1977). *J. Geophys.* **43**, 215–226.

Pratt, D. D., Moore, C. B., Parsons, M. L., and Anderson, D. L. (1977). *Proc. Lunar Conf. 8th, Geochim. Cosmochim. Acta, Suppl. 8* **2**, 1839–1847.

Pratt, D. D., Moore, C. B., and Parsons, M. L. (1978). Proceedings of the Ninth Lunar Science Conference, *Geochim. Cosmochim. Acta, Suppl. 10* **1**, 487–494.

Pratt, D. D., Moore, C. B., Parsons, M. L., and Anderson, D. L. (1978). *Res. Dev.* **2**, 53–64.

Puzyrev, N. N., and Gol'din, S. V. (1975). *Institut. Geologii i Geofiziki*, 215 pp.

Renoux, A., Butor, J. F., and Madelaine, G. (1975). *Chemosphere* **4**(3), 145–150.

Rhodes, J. M. (1969). *Lithos* **2**, 223–237.

Rutovitz, D. (1972). *Proc. Conf. on Machine Perception of Patterns and Pictures*, Teddington, Middlesex, England, pp. 81–87.

Saager, R., and Sinclair, A. J. (1974). *Mineral. Deposita (Berl.)* **9**, 243–252.

Salkie, M. L., and Luetchford, M. J. (1976). *Clin. Biochem.* **9**(5), 229–233.

Sasaki, S., and Ishida, Y. (1972). *Bunseki Kagaku* **21**, 1029–1037.

Schecter, J., and Jurs, P. C. (1973). *Appl. Spectrosc.* **27**, 225–232.

Shoenfeld, P. S., and DeVoe, J. R. (1976). *Anal. Chem.* **48**, 403R–411R.

Siromoney, R., Krithivasan, K., and Siromoney, G. (1973). *Proc. Indian Acad. Sci., Sect. A.* **78**, 72–88.

Smith, D. H., Buchanan, B. G. Englemore, R. S., Duffield, A. M., Yeo, A., Feigenbaum, E. A., Lederberg, J., and Djerassi, C. (1972). *J. Am. Chem. Soc.* **94**, 5962–5971.

Smith D. H., Buchanan, B. G., Englemore, R. S., Adlercreutz, H., and Djerassi, C. (1973) *J. Am. Chem. Soc.* **95**, 6078–6084.

Sotzberg, L. J., and Wilkins, C. L., (1976). *J. Am. Chem. Soc.* **98**, 4006.

Stark, H., and Lee, D. (1976). *IEEE Trans. Syst., Man, and Cybern.* **SMC-16**(11), 788–793.

Stuper, A. J., and Jurs, P. C. (1975). *J. Am. Chem. Soc.* **97**, 182–187.

Suenaga, Y., Toriwaki, J., and Fukumura, T. (1974). *Syst. Comput. Control* **5**(3), 35–43.

Tebbe, D. L. (1976). *Proc. of 1976 IEEE SE Region 3 Conf. on Eng. in a Changing Economy*, Clemson, South Carolina, pp. 199–200.

Thomas, Q. V., and Perone, S. P. (1977). *Anal. Chem.* **49**, 1369–1375.

Ting, K. L. H., Lee, R. C. T., Chang, C. L., and Guarino, A. M. (1975). *Comput. Biol. Med.* **4**, 301–332.

Tsukada, M., Ishii, N., and Sato, R. (1975). *Biol. Cybern.* **17**, 19–28.

Vandeginste, B. G. M., and Van Lersel, P. B. W. (1978). *Proc. Anal. Div. Chem. Soc.* **15**, 10–13.

Varma, O. P., and Jaipuriar, A. M. (1977). *Abstr. of 2nd Symposium on the Origin and Dist.* No. 77-0044, p. 163.

Videau, H. (1972). Report available from Dep. NTIS, FRNC-TH-372, 224 pp. [cited in *Nucl. Sci. Abstr.* (1973). **28**(8), 18368].

Vilela, J. A., Flach, G., Atmar, W., Ellington, J., and Pooler, J. (1976). *1976 Region IV IEEE Conf. Digest on Elect. Eng. for this Decade*, Austin, Texas, pp. 11–14.

Wagner, H. N., Jr., and Natarajan, T. K. (1972). *Digest of 3rd Int. Conf. on Med. Physics*, Gothenburg, Sweden, p. 36.2.

Webb, J., Kirk, K. A., Jackson, D. H., Niedermeier, W., Turner, M. E., Rackley, C. E., and Russel, R. O. (1976a). *Exp. Mol. Pathol.* **25**, 322–331.

Webb, J., Kirk, K. A., Niedermeier, W., Griggs, J. H., Turner, M. E., and James, T. N. (1976b). *Bioinorg. Chem.* **5**, 261–266.

Webb, J., Niedermeier, W., Griggs, J. H., Kirk, K. A., Turner, M. E., and James, T. N. (1976c). *Anal. Chim. Acta* **81**, 143–148.

Weiner, P. H., Liao, H. L., and Karger, B. L. (1974). *Anal. Chem.* **46**, 2182–2190.

Whitehouse, R. N. H. (1969). *Eucarpia Congr. Assoc. Eur. Amelior Plant* **5,** 61–96.

Wiatr, I., and Stenzel, P. (1974). *J. Int. Assoc. Math. Geol.* **6,** 17–31.

Wiatr, I., Stenzel, P., and Kowalski, A. (1974). *Biul. Geol. Warsaw Univ.* **16,** 35–66.

Wilkins, C. L., and Isenhour, T. L. (1975). *Anal. Chem.* **47,** 1849–1851.

Williams, R. D. (1976). *Proc. of Conf. on Appl. of Electronics in Medicine,* Southampton, England, IEEE, New York, pp. 179–186.

Winkel, P. and Statland, B. E. (1975). *Ann. Biol. Chim.* **33,** 174–182.

Wong, A. K. C., Vagnucci, A. H., and Liu, T. S. (1976). *IEEE Trans. Syst., Man, Cybern.* **SMC-6**(1), 33–45.

Yockey, H. P. (1973). *Biogenesis, Evolution, Homeostasis. A Symp. by Correspondence,* Springer-Verlag, New York, pp. 9–23.

Young, I. T., Walker, J. E., and Bowie, J. E. (1974). *IEEE Trans. on Inf. and Control* **25**(4), 357–370.

Zander, G. S., and Jurs, P. C. (1975). *Anal. Chem.* **47,** 1562–1573.

APPENDIX I
PATTERN RECOGNITION
DEFINITIONS AND REFERENCE
BOOKS

A. PATTERN RECOGNITION DEFINITIONS

Case: A sample that is analyzed for the set of variables in the data list (also called *object*).

Category: The division into groups, each of which contains patterns with similar characteristics (also called *groups*).

Classification: The determination of how the patterns are stratified into groups or categories.

Data vector: See Pattern.

Dependent variable: The variable to be calculated or predicted from a set of other variables.

Feature: A variable that has been changed or transformed by some algorithm.

Groups: See Category.

Independent variable: One variable in the set of variables used to predict or calculate the value of a dependent variable.

Multivariate analysis: Manipulations that consider more than one variable simultaneously.

Nonparametric statistics: Statistical techniques that do not depend upon a specific parent distribution.

Normal distribution: See Appendix II.

Object: See Case.

Parametric statistics: Statistical techniques that rely on the validity of the underlying distributional assumptions (usually Gaussian).

Pattern: A complete list of features and/or variables used to characterize a given case (also called *data vector*).

Pattern recognition: The study of data sets in order to find regularities and similarities inherent in the data.

Scaling: See Appendix IX (also called *standard scores*).

Similarity: Some measure of how closely two patterns resemble each other. Many algorithms are available to define this (also called *measure of association*).

Standard scores: See Appendix IX.

Supervised learning: Methods employed to check known stratified groups as to the validity of the grouping (or classification).

Test set: The set of patterns utilized after classification has been defined. The goal of this analysis (called the *test step*) is to predict classification for these patterns.

Training set: The set of patterns used as the initial set when the goal is to predict or define classification. This step is called the *training step*.

Unsupervised learning: Using patterns with no *a priori* stratification to study "natural" groupings or classifications.

Variable: A given characteristic that is measured for the cases included in the data set (also called *measurement*).

NOTE: See the index for references to material in this book further defining these terms. Descriptions of statistical terms should be sought in one of the statistical analysis books referenced below.

B. REFERENCE BOOKS

Pattern Recognition and Multivariate Analysis (Mathematical):

H. C. Andrews, *Introduction to Mathematical Techniques in Pattern Recognition,* John Wiley and Sons, New York, 1972.

R. O. Duda and P. E. Hart, *Pattern Classification and Scene Analysis,* John Wiley and Sons, New York, 1973.

K. Fukunaga, *Introduction to Statistical Pattern Recognition,* Academic, New York, 1972.

R. G. Gnanadesikar, *Methods for Statistical Data Analysis of Multivariate Observations,* John Wiley and Sons, New York, 1977.

D. F. Morrison, *Multivariate Statistical Methods,* 2nd Ed., McGraw-Hill, New York, 1976.

N. J. Nilsson, *Learning Machines,* McGraw-Hill, New York, 1965.

Pattern Recognition and Multivariate Analysis (Applied):

S. Bennett, *An Introduction to Multivariate Techniques for the Social and Behavioral Sciences,* John Wiley and Sons, New York, 1976.

P. R. Bevington, *Data Reduction and Error Analysis for the Physical Sciences,* McGraw-Hill, New York, 1969.

P. C. Jurs and T. L. Isenhour, *Chemical Applications of Pattern Recognition,* Wiley-Interscience, New York, 1975.

G. S. Koch, Jr., and R. F. Link, *Statistical Analysis of Geological Data,* John Wiley and Sons, New York, 1970.

B. R. Kowalski, *Chemometrics: Theory and Application,* American Chemical Symposium Series #52, 1977.

P. M. Mather, *Computational Methods of Multivariate Analysis in Physical Geography,* John Wiley and Sons, New York, 1976.

H. Saai, *Multivariate Statistical Analysis for Biologists,* Methuen, London, 1964.

Specific Techniques:

B. Everitt, *Cluster Analysis,* Heinemann Educational Books, London, 1974.

H. R. Harman, *Modern Factor Analysis,* 3rd Ed., University of Chicago Press, 1980.

P. A. Lachenbruck, *Discriminant Analysis,* Hafner Press, 1975.

E. R. Malinowski and D. G. Howery, *Factor Analysis in Chemistry,* John Wiley and Sons, New York, 1980.

J. Neter and W. Wasserman, *Applied Linear Statistical Models,* Irwin, Inc., 1974.

M. L. Puri and P. K. Sen, *Nonparametric Methods in Multivariate Analysis,* John Wiley and Sons, New York, 1970.

Basic Statistical Analysis Books:

G. E. P. Box, W. G. Hunter, and J. S. Hunter, *Statistics for Experimentalists: An Introduction to Design, Data Analysis, and Model Building,* John Wiley and Sons, New York, 1978.

B. S. Everitt, *Graphical Techniques for Multivariate Data,* Heinemann Educational Books, London, 1974.

W. L. Hays, *Statistics for the Social Sciences,* Holt, Rinehart, and Winston, New York, 1973.

R. F. Hirsch, ed., *Statistics,* The Franklin Institute Press, 1978.

M. Lentner, *Introduction to Applied Statistics,* Prindle, Weber, and Schmidt, Inc., 1975.

S. J. Press, *Applied Multivariate Analysis,* Holt, Rinehart, and Winston, New York, 1972.

A. Romano, *Applied Statistics for Science and Industry,* Allyn and Bacon, Inc., Boston, 1977.

R. G. D. Steel and J. H. Torrie, *Principles and Procedures of Statistics,* McGraw-Hill, New York, 1960.

APPENDIX II
THE MULTIVARIATE NORMAL
DISTRIBUTION

Much of parametric statistical analysis is based on the assumption that the data follow a normal or Gaussian distribution, an example of which is shown in Table A.II. Given the measurements of a certain variable on an infinite number of cases, a histogram of the values should take a form similar to that shown. The distribution is symmetrical around a single peak value. This value is represented as the mean value (symbolized as μ). It is also assumed that the density falls off steadily in each direction from this point, although it never theoretically reaches zero. A standard deviation (σ) is used to designate the abruptness of this decrease. The decrease in the height of the curve accelerates and the curve is convex until a point one standard deviation unit above or below the mean is reached. At this point, the decrease decelerates and the curve becomes concave. In practice, measurements can never be made on an infinite number of cases. A number of parametric statistical methods will be approximately valid if the central limit theorem holds.

This theorem states that if a variable (x) has a distribution of values with a standard deviation of σ and a mean of μ and if a sample size of N is taken, then the distribution of the sample mean $\bar{x}$, will approach a normal Gaussian shape as N approaches

TABLE A.II
Number of Samples Having a Given Value

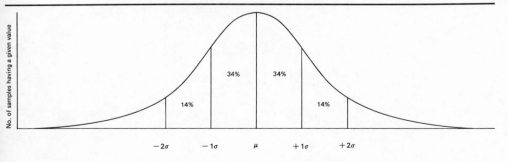

infinity. In reality, distributions that are not exactly symmetrical (called skewed) can still be used if the deviations from Gaussian shape are not great. Adherence of the data to the normal distribution can be checked in Step II.1d. If adherence is not good, then nonparametric statistics (see Appendix VII) should be considered.

In the multivariate case, where many variables are considered simultaneously, each variable is assumed to be normally distributed. This means that the distribution is symmetrical with a single peak value occurring at the point where all variables take on their mean value. It is also assumed that any linear combination of variables will also be normally distributed.

The multivariate central limit theorem is useful in many cases. This theorem states that if variables $x_1, x_2, \ldots, x_p$ have standard deviations $\sigma_1, \sigma_2, \ldots, \sigma_p$ and correlations ρ_{ij}, where $i = 1 \cdots p$ and $j = 1 \cdots p$, then the means $\mu_1 \mu_2, \ldots, \mu_p$ of the sample of size N have an approximate joint multivariate normal distribution.

APPENDIX III
DATA BASE DESCRIPTION

PART A. HOW TO PRESENT DATA FOR COMPUTER
COMPATIBILITY

When the format for a data set is prepared for the types of analyses in this book, it is strongly advised to use what we refer to as an "ARTHUR-type" format. Of the three programs discussed here, ARTHUR has, by far, the most restrictions on the acceptable data format. Therefore, if the data are coded according to these rules, it will also be compatible with both SPSS and BMDP. There are many options for doing this. We will limit this discussion to the simplest and most convenient method for entering the data into the computer. Other frills may be added later. ARTHUR requires a separate card (or set of cards) for each pattern, containing the pattern identification and variable values, utilizing the format

$$(\text{I}i, \text{A}j_1, \text{A}j_2, \text{F}a.b, n\text{F}c.d)$$

where i is the number of columns used to define the identification, j_1 is the number of columns used for the first word of the name, j_2 is the number of columns used for the second word of the name, a.b is a floating point descriptor for the category number, n is the number of variables, and c.d is a floating point descriptor for the variable format. Translating this into English, there are three types of formats with which we are concerned. An "I-type" format permits only whole number integers to be used. An "A-type" format reads letters or numbers but considers the symbol to be a name and not a measureable value on which statistics can be performed. An "F-type" format is used to indicate either integers or decimal numbers which can be used in mathematical calculations. The user must realize that a computer card consists of 80 columns (or writing spaces) and a format statement is used to indicate to the computer in which columns a given variable is found. Therefore, using an AUTHUR-type format, the following example should be studied.

If the program is given the following format:

(I1,A3,A4,F2.0,6F5.3)

this means

I1: One column is used for a one-digit identification number 0–9.
A3: The first word of the name takes three spaces.
A4: The second word of the name takes four spaces.
F2.0: The category number takes two spaces (the ".0" means to except no digits after the decimal point).
6F5.3: The six variables to be considered each take five columns and the last three places are considered to fall after a decimal point.

The computer would then expect input of the form

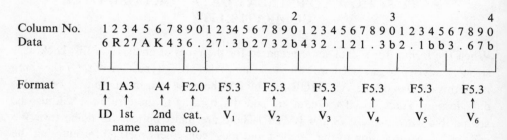

(Note that in the above text, a "b" is used to symbolize a blank space on the card.) The ID is 6 (for the user to remember it is the sixth case or something meaningful like that). The two names of this case are R27 and AK43 and it belongs to the sixth category. From here, the computer counts off blocks of five columns for the six requested variables (designated here as V1 to V6). The user can name these later. In the "F a.b" format, if a decimal point is present, the computer recognizes it. If one is not, the computer puts one in the bth place from the right end of the number. It first reads 27.3. It then encounters "2732b" which are the next five columns. It reads the numbers as "27320" because blanks are equivalent to zeros. Since no decimal point is present, the computer inserts one to follow F5.3 format; therefore, it reads the value as 27.320. It then assigns V3 to V6 values of 432.1, 21.3, 2.1, and 3.67, respectively. More practice in this can be obtained in an introductory FORTRAN manual.

A few additional points should be considered. Each variable does not have to take up the same number of spaces. Assume the first takes up 6, the next two 7, the fourth 3, and the fifth and sixth 5 each. The format could be

(Ii,Aj₁,Aj₂,Fa.b,F6.d,2F7.e,F3.f,2F5.g).

In this format, the letters d, e, f, and g are used to designate the number of decimal places. Note that the number in front of the F indicates how many times that format is repeated.

A second point to be made is that it is much easier, especially for beginners, always to include the decimal points. Everything can then be assigned an Fa.0 format since the computer ignores the ".0" if a decimal point is included. This eliminates the problems of "2732b" being read as 27320 or 27.320 depending on whether F5.0 or F5.3 is specified.

Cases can be continued on a second card, if necessary. To include this in a format statement, a "/" symbol is used. The computer will then go to the next card regardless of whether all of the data on that card have been read or not (it only reads the columns the user specifies in the format statement, regardless of what else is present on the card). An example is (I1, 2A3, F2.0, F3.0, F4.0/2F3.0). This reads the ID in column 1, the two-word name in columns 2–4 and 5–7, the category in columns 8–9, the first variable in columns 10–12 of card one, the second in columns 13–16, and the third and fourth in columns 1–3 and 4–6 of the second card. Any other data present on the two cards will be ignored.

For ARTHUR, there must be at least one column assigned to the ID. The ID is not used by the program, but can be useful to the user for marking cards. It is often left blank since ARTHUR always assigns and uses its own case ID.

ARTHUR requires at least two names of at least one column each. The name can be any length. However, in the subroutines, if the user specifies the name to be plotted (such as in VARVAR) only the first two characters of each name will be used due to the limited space. Therefore, either names should be limited to four symbols (2A2 format) or at least the first two symbols of each name should be the most important to retain their meaning when plotted.

Categories may not be used in every statistical calculation. However, be sure to leave enough room in the category designation since they may be included at a later time. More than one card can be used per case, and it is advisable to spread the data out on the cards with spaces between analyses so that it will be easier to read. In the format, these spaces can be designated in two ways. Since blank spaces are considered as zeros to the computer, a nine-column readout of bbb2.34bb is read as 002.3400 or given the value of 2.34. This value, read in F9.0 format, is still 2.34 but allows blanks on the card between analyses for easier reading. A second way to do this is to use a fourth type of format designation, aX, which indicates to the computer to ignore "a" number of columns regardless of what is present in them. This is another way to ignore data present on the card that you do not want to consider. Therefore, bbb2.34bb can be formatted as (3X,F4.0,2X).

The data format should always be checked carefully for mistakes. One common mistake is to omit the parentheses around the format statement. Unfortunately ARTHUR is not much help in identifying errors. The first few steps of II.1 should be a very helpful aid to correct format errors using SPSS.

For the air pollution data listed below, the format used was made ARTHUR compatible. Ten variables were used in the data base. The following format was specified:

(I1,2A4,F4.0,2F5.0,2X,6F7.0,F5.0,F6.0)

Where

> I1: ID number (ARTHUR assigns his own ID number; therefore, Column 1 was left blank on all cards).
>
> 2A4: Two four-letter names. These are coded to indicate the place and date. [For instance, in the first case, MC75JR12 indicates the data were obtained from MC (Maricopa County) in 1975 on January 12th.]
>
> F4.0: Four spaces for category designation. This was left blank in the original data. Later, it was filled in with
> 1.0 = MC (Maricopa County)
> 2.0 = MS (Mesa)
> 3.0 = NS (North Scottsdale)
> 4.0 = SC (Scottsdale)
>
> 2F5.0: The first two variables, MN and CU, took up five spaces each. A decimal was included in the data itself.
>
> 2X: Two ignored spaces. A zero or one was used here to indicate the presence or absence of CD for later use. However, in the first analyses, these columns were ignored.
>
> 6F7.0: Six variables, FE, PB, PART, ORG, SULF, and NIT took seven columns each for data.
>
> F5.0: Variable number nine (CL) took five columns.
>
> F6.0: Variable number ten (CO) took six columns.

For analyses whose values were not available, a missing value designation had to be used (see Appendix VIII). Normally in ARTHUR, this is designated by filling the entire format area for that variable in that case with 9's (see an example in the sixth case in Table A.III).

PART B. METHODS OF ANALYSIS AND RAW DATA LISTING

The following standard County Health Department methods of analyses were used to obtain data for the air pollution base. All are reported in units of micrograms per meter cubed of air.

Particulates (PART)

Particulate concentrations were measured by passing a metered flow of air for 24 hours through a preweighed 8 × 10-in. glass fiber filter. Particulates in the air sample were trapped on the filter, and the gain in weight of this was used. The concentration was expressed as weight of particulates per volume of air passed.

This filter was divided into sections for the analysis of benzene-soluble organics (ORG), sulfates (SULF), nitrates (NIT), lead (PB), and other metals.

Benzene Soluble Organics (ORG)

The particulate samples were extracted in a hot benzene solution. The majority of the species found were from incomplete combustion of fuels that were released into the atmosphere.

Nitrates (NIT)

Particulate samples were extracted into water. Colorimetric analyses were used to determine nitrate concentrations.

Sulfates (SULF)

Particulate samples were extracted into water. Turbidimetric analyses were used to determine sulfate concentrations.

Lead (PB)

Particulate samples were extracted into nitric acid. Atomic absorption analyses were used to determine the lead concentrations.

Other Metals (MN, CT, CD, FE)

Acid extracts were analyzed by atomic absorption methods, similar to the lead analyses.

Carbon Monoxide (CO)

Two methods were used to analyze for CO. Direct infrared absorption analyses were performed. Gas chromatographic techniques were also used to separate CO from the other gases present. The CO effluent from the column was converted to methane, which was subsequently measured in a flame ionization detector.

Chlorides (CL)

An acid extract of the particulate matter samples was analyzed for chlorides by using a Ag^+ titration.

TABLE A.III

NAME	CAT MN	CU	CD	FE	PB	PART	ORG	SULF	NIT	CL	CO
MC75JR12	1.0.04	2.96	0	1.35	2.06	79.66	11.00	2.55	0.83	1.39	1814.
MC75JR18	1.0.09	3.11	1	2.63	5.61	169.07	18.10	3.06	4.39	9.33	7303.
MC75JR24	1.0.12	1.68	1	3.79	4.84	197.43	14.84	6.37	4.31	8.07	7351.
MC75JR30	1.0.06	2.05	1	1.83	0.92	93.78	2.29	4.18	1.96	4.01	1289.
MC75FR05	1.0.31	2.01	0	1.57	2.36	86.91	10.47	2.73	4.03	3.19	3675.
MC75FR11	1.0.06	1.88	0	2.55	3.73	131.19	12.21	99999994.90		999994487.	
MC75FB23	1.0.04	0.96	0	1.63	3.07	97.02	10.86	2.58	2.38	5.38	2816.
MC75MR01	1.0.13	1.12	0	3.37	4.12	204.84	14.03	7.49	4.00	9.56	3580.
MC75MR07	1.0.10	1.05	0	2.55	2.94	156.27	8.83	5.05	2.86	8.77	3866.
MC75MR13	1.0.03	1.10	0	0.75	1.82	114.99	4.39	4.42	4.41	3.52	2482.
MC75MR19	1.0.01	0.0	0	0.0	0.0	115.35	6.37	6.71	9999999.52		1527.
MC75MR25	1.0.12	0.85	0	3.52	0.90	190.83	5.87	4.38	3.08	3.03	286.
MC75MR31	1.0.13	0.71	0	3.63	0.88	173.99	1.25	4.95	2.06	2.71	1193.
MC75AP24	1.0.05	1.15	0	1.64	0.61	87.73	5.34	2.91	2.38	1.66	1464.
MC75AP30	1.0.09	0.90	0	2.57	2.17	129.05	5.64	3.69	1.81	3.37	3103.
MC75MY06	1.0.38	0.84	0	2.67	1.30	125.86	3.20	0.65	0.74	2.35	1718.
MC75MY30	1.0.06	0.93	0	1.72	1.12	98.90	3.42	0.81	1.21	0.87	1384.
MC75JN17	1.0.09	0.53	0	2.91	0.34	143.20	0.0	2.12	0.20	4.76	999999
MC75JN23	1.0.06	0.75	0	2.62	0.79	97.44	1.20	2.11	1.63	3.64	999999
MC75JN29	1.0.06	1.01	0	2.49	1.24	99.67	6.11	2.66	2.87	3.85	1432.
MC75JL05	1.0.04	1.69	0	1.46	0.40	88.64	4.19	5.14	3.26	5.84	239.
MC75JL29	1.0.09	1.57	0	3.14	0.52	84.35	0.22	4.89	3.15	8.50	1337.
MC75AG04	1.0.05	1.24	0	1.82	1.19	93.63	3.63	6.79	2.86	8.24	1400.
MC75AG10	1.0.17	1.06	0	5.64	0.76	287.02	7.73	7.50	4.42	3.31	1718.
MC75AG16	1.0.04	0.99	0	1.57	0.82	86.68	3.76	4.72	3.67	6.45	999999
MC75AG22	1.0.07	1.01	0	2.18	2.05	114.77	4.22	4.12	3.14	2.67	999999
MC75AG28	1.0.08	1.30	0	2.34	1.56	129.15	3.83	5.01	1.96	4.75	999999
MC75SP03	1.0.07	0.92	0	1.92	0.99	102.96	3.83	10.01	2.97	2.90	999999
MC75SP09	1.0.04	1.25	0	1.18	0.90	81.93	2.36	5.81	2.70	1.19	999999
MC75SP15	1.0.05	1.24	0	1.51	2.51	92.04	99999995.58		2.55	4.81	999999
MC75SP21	1.0.04	1.01	0	1.36	1.12	74.54	1.98	2.52	1.56	0.91	999999
MC75SP27	1.0.06	1.00	0	1.77	1.88	95.49	4.36	5.35	1.75	2.84	4153.
MC75OC03	1.0.08	1.02	0	1.99	2.18	121.41	4.64	4.36	0.17	0.97	4869.
MC75OC09	1.0.10	0.77	1	3.05	2.22	139.11	4.17	4.73	2.21	5.06	4773.
MC75OC21	1.0.02	0.61	0	0.26	0.40	27.47	1.71	2.36	0.59	0.99	2005.
MC75OC27	1.0.12	1.10	0	1.85	2.46	129.15	8.40	7.45	1.19	3.67	5012.
MC75NV14	1.0.09	1.01	0	3.05	4.12	183.56	27.0	14.98	0.33	4.87	16337.
MC75DC08	1.0.05	0.72	0	1.57	4.02	126.53	18.74	0.0	0.0	0.0	10310.
MC75DC20	1.0.02	0.93	0	0.33	1.90	66.09	11.90	6.85	2.34	2.09	999999
MC75DC26	1.0.33	1.52	0	1.67	3.43	108.55	20.65	2.59	1.99	2.48	8926.
MC76JR13	1.0.10	1.90	0	3.09	4.36	211.91	20.99	7.18	7.47	4.30	8783.
MC76JR19	1.0.10	1.58	1	2.60	3.68	173.72	12.14	7.35	1.00	13.9	6923.
MC76FB06	1.0.01	2.18	0	0.94	1.58	65.84	5.85	4.17	1.86	0.78	3962.
MC76FB18	1.0.05	3.38	0	2.24	1.88	117.48	7.46	2.96	6.53	2.11	1146.
MC76FB24	1.0.08	2.22	0	3.34	3.20	146.48	11.17	3.44	1.61	4.56	999999
MC76MR01	1.0.07	3.05	0	3.02	0.88	142.09	3.86	2.43	2.28	2.87	3600.
MC76MR07	1.0.03	3.82	0	1.22	1.33	86.78	6.82	3.52	2.80	2.03	3437.
MC76MR13	1.0.06	2.68	0	2.30	3.79	146.53	12.11	3.32	1.47	3.79	4732.
MC76MR19	1.0.11	2.96	0	3.18	1.46	187.40	5.17	4.37	3.92	4.25	2959.
MC76MR25	1.0.12	2.31	0	3.52	1.08	206.35	2.89	3.49	3.15	2.84	2721.
MC76MR31	1.0.09	2.74	0	2.77	3.32	160.99	8.15	3.68	2.78	2.35	3389.
MC76AP12	1.0.07	2.68	0	2.33	0.55	140.80	3.58	3.40	2.36	4.51	191.
MC76AP24	1.0.06	2.74	0	2.17	2.39	122.29	6.85	3.14	3.60	4.46	1957.
MC76MY06	1.0.05	5.23	1	1.66	0.61	110.48	1.42	4.28	1.97	3.42	382.
MC76MY12	1.0.07	3.61	0	1.90	1.12	116.74	3.62	4.20	3.50	1.16	430.
MC76MY18	1.0.07	4.35	0	1.86	0.52	112.04	2.43	8.48	3.05	1.34	191.
MC76MY24	1.0.06	999990		1.67	0.64	120.75	3.01	2.82	1.48	2.48	1193.
MC76JN05	1.0.06	0.10	1	1.88	0.35	134.47	0.76	2.57	2.23	4.28	1766.
MC76JN11	1.0.06	0.10	0	1.99	0.72	120.20	2.89	2.70	0.38	2.94	1241.
MC76JN23	1.0.06	0.19	0	1.45	0.44	73.15	2.13	2.72	1.90	0.75	1098.
MC76JL05	1.0.06	0.08	0	2.06	0.23	94.32	2.76	4.89	1.25	1.78	382.
MC76JL17	1.0.03	0.05	0	1.12	0.10	57.72	1.34	4.40	1.61	1.31	0.0
MC76JL23	1.0.06	0.10	0	2.02	0.63	113.79	3.31	10.25	5.84	0.97	573.
MC76JL29	1.0.07	0.06	0	2.36	0.54	123.30	0.99	6.25	3.37	1.94	999999
MC76AG10	1.0.12	0.07	0	2.98	0.67	186.22	3.43	6.25	2.17	4.46	674.
MC76AG16	1.0.09	0.08	0	3.21	1.82	166.76	4.67	5.77	1.18	3.15	859.
MC76AG22	1.0.04	0.05	0	1.45	0.62	84.18	2.18	5.92	1.95	2.08	0.0
MC76SP27	1.0.04	0.07	0	1.26	1.51	80.70	3.62	3.15	0.91	1.58	1074.
MC76OC03	1.0.05	0.07	1	2.47	2.47	132.48	5.09	5.19	2.53	1.89	1957.
MC76OC21	1.0.06	4.02	0	1.59	2.96	107.98	10.03	10.76	1.17	4.34	3628.
MC76OC27	1.0.02	1.89	0	0.98	1.02	67.20	3.09	4.27	0.91	3.76	2673.
MC76NV02	1.0.06	2.87	0	2.30	6.59	137.01	14.45	7.62	0.86	6.81	7446.
MC76NV08	1.0.07	2.05	0	2.73	5.92	155.94	13.50	9.51	0.53	7.16	8087.
MC76NV14	1.0.03	2.48	0	0.81	1.77	79.84	8.55	7.14	2.71	0.36	1432.
MC76NV20	1.0.06	2.89	0	2.19	5.02	144.69	12.91	10.18	8.48	3.92	6110.
MC76NV26	1.0.07	2.40	0	2.30	2.75	134.03	7.80	8.73	3.43	5.86	3150.
MC76DC08	1.0.16	4.42	0	5.61	7.78	283.94	28.48	6.85	4.60	11.1	8687.
MC76DC14	1.0.15	5.14	0	4.50	7.77	222.29	23.54	11.49	3.00	13.2	8401.
MC76DC20	1.0.05	99999952.12			2.05	114.72	5.91	13.66	2.20	3.83	1432.
MC77JR07	1.0.08	3.13	0	3.36	4.20	146.52	19.41	5.84	3.47	7.11	4153.
MC77FB18	1.0.16	0.27	1	99999994.30		253.94	14.68	8.56	1.67	5.93	5680.

TABLE A.III (*Continued*)

NAME	CAT	MN	CU	CD	FE	PB	PART	ORG	SULF	NIT	CL	CO
MC77FB24	1.0	0.12	0.15	0	99999	991.40	186.31	4.42	2.73	0.45	7.12	1814.
MC77MR06	1.0	0.17	0.54	0	5.34	4.08	246.35	13.27	7.29	1.52	6.73	5537.
MC77MR14	1.0	0.11	0.08	0	3.90	1.66	165.50	6.81	4.13	3.58	3.53	1326.
MC77MR20	1.0	0.11	0.07	0	4.17	2.40	177.10	6.75	3.89	0.19	4.36	3103.
MC77MR26	1.0	0.04	0.07	0	0.97	0.99	52.00	2.70	2.57	0.79	2.33	1289.
MC77AP01	1.0	0.07	0.06	0	3.08	0.90	127.56	1.64	4.76	4.38	3.91	1504.
MC77AP07	1.0	0.12	0.15	0	4.78	4.05	201.99	13.08	12.91	1.66	5.33	5728.
MC77MY01	1.0	0.09	0.06	0	99999	991.00	147.64	4.97	3.83	1.29	8.34	3532.
MC77MY07	1.0	0.10	0.04	0	99999	991.00	169.68	0.58	3.15	0.63	13.0	1146.
MC77MY13	1.0	0.04	0.04	1	99999	991.12	70.38	1.12	3.99	1.10	3.83	1527.
MC77MY19	1.0	0.12	0.44	1	99999	992.83	192.57	6.69	4.39	3.31	5.97	2760.
MC77JN06	1.0	0.07	0.08	0	2.99	0.90	141.73	3.05	14.53	1.90	5.45	1575.
MC77JN18	1.0	0.11	0.06	0	4.40	1.55	192.00	1.75	4.85	1.22	6.00	1909.
MC77JN24	1.0	0.15	0.07	0	5.22	1.26	214.44	1.10	5.58	2.20	4.61	1814.
MC77JL12	1.0	0.10	0.04	1	2.86	0.80	173.62	3.27	6.47	3.00	4.28	99999
MC77JL30	1.0	0.05	0.04	0	1.81	0.32	104.31	0.89	3.66	1.83	5.88	1241.
MC77AG05	1.0	0.06	0.06	0	2.35	1.28	118.32	2.62	3.62	2.73	5.10	2540.
MC77AG11	1.0	0.05	0.05	0	2.05	0.59	98.58	0.85	4.75	1.40	1.39	1550.
MC77AG17	1.0	0.03	0.05	0	0.92	0.38	52.43	0.09	3.30	3.41	2.98	1241.
MC77AG23	1.0	0.03	0.07	0	1.08	0.42	67.79	0.28	2.90	2.76	3.44	1146.
MC77AG29	1.0	0.10	0.07	0	3.92	2.37	168.69	5.59	3.43	0.16	3.46	4201.
MC77SP04	1.0	0.07	0.10	0	2.98	1.60	128.23	2.82	11.18	0.09	2.72	3055.
MC77SP10	1.0	0.07	0.06	0	2.19	1.07	108.85	0.46	5.47	0.09	2.11	1241.
MC77SP16	1.0	0.06	0.08	0	5.60	0.90	109.13	1.18	3.74	0.06	4.41	1814.
MC77OC04	1.0	0.07	0.10	0	5.72	2.31	118.90	4.39	9.34	0.0	0.68	2673.
MC77OC10	1.0	0.05	0.05	0	2.11	2.01	93.64	5.46	5.50	0.0	1.05	5728.
MC77OC28	1.0	0.07	0.06	0	6.40	1.41	120.26	2.83	4.77	0.09	3.71	2434.
MC77NV03	1.0	0.08	0.19	0	7.92	4.83	151.12	10.15	10.94	0.08	2.65	6062.
MC77NV09	1.0	0.09	0.14	0	4.12	6.08	143.91	14.86	4.99	0.10	3.72	6826.
MC77NV15	1.0	0.13	0.13	0	5.12	5.70	186.41	12.58	8.22	1.63	5.96	7208.
MC77NV21	1.0	0.12	0.08	0	4.89	5.51	195.41	20.51	7.34	0.47	8.64	7222.
MC77NV27	1.0	0.11	0.11	0	4.28	3.10	155.63	9.82	8.68	2.43	3.90	3580.
MC77OC03	1.0	0.15	0.16	0	6.48	5.61	220.80	16.43	6.78	1.42	0.0	5823.
MS75JR12	2.0	0.04	0.03	0	1.54	1.09	76.11	2.03	1.97	0.58	1.20	99999
MS75JR18	2.0	0.08	0.06	1	2.23	2.48	128.87	9.30	3.05	7.13	2.84	99999
MS75JR24	2.0	0.11	0.13	1	3.38	2.13	180.78	9.19	7.24	2.02	4.66	2291.
MS75JR30	2.0	0.06	0.09	0	1.94	0.94	94.31	7.83	2.08	0.51	2.92	1337.
MS75FB05	2.0	0.04	0.16	0	1.55	1.94	83.87	4.96	1.32	3.08	0.26	1623.
MS75FB11	2.0	0.07	0.12	0	2.52	2.27	119.20	5.72	0.83	2.12	0.0	99999
MS75FB23	2.0	0.05	0.18	0	1.88	2.14	94.76	5.50	1.41	3.12	2.22	99999
MS75MR01	2.0	0.06	0.26	0	2.51	2.33	112.08	6.80	2.94	3.12	2.70	99999
MS75MR07	2.0	0.12	0.18	0	3.14	0.39	234.73	3.38	3.56	1.93	3.85	99999
MS75MR13	2.0	0.03	0.07	0	0.91	0.58	62.41	1.08	3.65	2.18	1.53	99999
MS75MR19	2.0	0.07	0.07	0	1.90	0.92	116.41	8.72	1.64	5.96	0.48	99999
MS75MR25	2.0	0.11	0.07	0	3.47	0.39	199.45	0.0	4.18	99999	992.99	99999
MS75MR31	2.0	0.14	0.15	0	3.82	0.32	216.88	3.34	5.27	0.13	2.17	430.
MS75AP24	2.0	0.09	0.10	0	2.69	1.54	142.45	9.49	1.85	2.91	1.83	1909.
MS75AP30	2.0	0.08	0.06	0	2.50	1.28	134.08	3.73	0.67	2.29	2.20	2339.
MS75MY06	2.0	0.09	0.14	1	2.56	1.02	130.19	5.33	0.45	0.13	1.99	1384.
MS75MY30	2.0	0.05	0.05	0	1.55	0.84	83.06	1.94	0.12	0.32	0.53	1241.
MS75JN17	2.0	0.13	0.24	0	3.68	0.29	204.42	0.0	3.07	0.38	5.41	573.
MS75JN23	2.0	0.09	0.11	0	4.05	0.64	136.96	1.18	2.62	1.44	3.48	1813.
MS75JN29	2.0	0.08	0.11	0	2.87	0.08	112.39	3.83	1.60	2.36	2.29	1241.
MS75JL05	2.0	0.05	0.12	0	1.87	0.42	109.47	0.0	5.08	3.74	1.81	0.0
MS75JL29	2.0	0.10	3.02	0	2.77	0.50	120.34	0.0	3.49	1.87	1.04	859.
MS75AG04	2.0	0.05	2.31	0	1.89	1.19	110.05	3.64	7.14	0.77	5.50	3675.
MS75AG10	2.0	0.13	3.43	0	4.45	0.48	228.20	7.40	7.56	3.88	3.47	430.
MS75AG16	2.0	0.06	3.38	0	1.90	1.08	116.86	2.71	4.04	1.99	3.93	1384.
MS75AG22	2.0	0.05	3.22	0	1.90	0.95	124.48	2.71	4.33	2.45	1.55	1480.
MS75AG28	2.0	0.10	1.98	0	3.20	1.09	166.55	2.47	3.55	1.18	2.79	1862.
MS75SP03	2.0	0.10	1.88	0	2.84	0.80	158.36	8.44	8.82	1.10	2.61	811.
MS75SP09	2.0	0.03	1.69	0	1.14	0.85	89.95	8.37	9.04	2.41	4.46	1193.
MS75SP15	2.0	0.05	1.21	0	1.37	0.92	103.41	4.17	3.81	1.63	1.47	1862.
MS75SP21	2.0	0.05	1.08	0	1.36	0.50	82.53	0.72	3.33	0.94	0.94	143.
MS75SP27	2.0	0.08	0.85	0	2.43	1.66	132.51	4.99	7.12	1.75	2.60	2434.
MS75OC03	2.0	0.10	0.92	0	2.56	1.95	152.22	4.35	7.24	0.05	2.98	2148.
MS75OC09	2.0	0.05	0.87	1	3.10	1.40	156.47	99999	997.15	2.82	4.47	1480.
MS75OC21	2.0	0.04	1.12	0	0.80	0.91	75.26	4.81	5.45	0.86	1.89	621.
MS75OC27	2.0	0.06	1.77	0	1.44	1.19	111.86	4.92	7.52	0.42	2.98	2673.
MS75NV14	2.0	0.07	1.18	0	3.07	2.28	197.15	39.51	2.88	0.44	0.26	6973.
MS75DC08	2.0	0.07	0.41	0	2.01	1.53	120.97	12.02	3.77	1.90	1.41	3771.
MS75DC20	2.0	0.05	0.29	0	1.17	0.75	87.38	1.92	4.32	1.51	1.77	2673.
MS75DC26	2.0	0.04	0.35	0	1.20	1.59	87.28	7.33	1.99	1.07	0.83	3485.
MS76JP13	2.0	0.03	0.25	0	1.09	0.33	60.59	3.07	1.92	0.88	0.58	5585.
MS76JR19	2.0	0.07	0.34	0	2.24	0.49	133.06	99999	992.79	1.17	1.27	4153.
MS76FB06	2.0	0.01	0.70	1	0.45	0.29	28.97	0.87	3.20	0.90	2.03	239.
MS76FB18	2.0	0.04	0.81	0	1.27	0.59	76.99	2.94	2.90	4.21	0.67	2493.
MS76FB24	2.0	0.05	0.79	1	2.36	1.24	129.44	5.10	1.92	1.68	1.40	99999
MS76MR01	2.0	0.02	0.48	0	1.15	0.0	60.30	1.55	0.86	0.75	1.05	5346.
MS76MR07	2.0	0.02	0.81	0	0.81	0.32	52.60	0.98	3.07	1.27	0.77	334.

TABLE A.III (*Continued*)

NAME	CAT	MN	CU	CD	FE	PB	PART	ORG	SULF	NIT	CL	CO
MS76MR13	2	0.04	0.89	0	1.56	0.99	80.76	3.01	2.95	0.41	0.76	797.
MS76MR19	2	0.06	0.78	1	1.99	0.59	112.12	2.75	3.26	1.89	2.41	3103.
MS76MR25	2	0.07	0.76	0	2.34	0.48	135.20	2.38	1.92	2.12	1.70	3103.
MS76MR31	2	0.04	0.99	0	1.27	0.67	82.16	2.45	2.56	0.95	2.86	2482.
MS76AP12	2	0.06	0.86	0	2.25	0.38	130.85	3.46	2.43	0.35	4.77	2721.
MS76AP24	2	0.03	0.88	0	1.27	0.84	81.67	2.60	2.30	2.10	1.65	999999
MS76MY06	2	0.02	0.85	0	0.86	0.32	58.55	0.78	3.22	2.02	4.12	999999
MS76MY12	2	0.04	1.08	0	1.18	0.35	80.32	0.39	1.75	1.91	0.90	1064.
MS76MY18	2	0.06	1.32	0	1.80	0.28	126.20	3.94	8.37	1.89	1.65	1098.
MS76MY24	2	0.06	1.30	0	1.56	0.45	102.34	1.60	2.59	0.92	2.13	95.
MS76JN05	2	0.05	1.10	1	1.42	1.03	94.15	4.41	1.72	1.24	3.89	1480.
MS76JN11	2	0.06	0.98	1	2.11	0.53	125.66	999999	1.84	0.53	3.61	1909.
MS76JN23	2	0.07	1.12	0	1.95	0.52	111.89	3.12	1.84	0.56	0.06	573.
MS76JL05	2	0.08	0.81	0	2.89	0.31	140.24	999999	5.90	0.58	2.66	999999
MS76JL17	2	0.04	0.69	0	1.50	0.23	86.10	2.43	5.14	2.46	2.54	0.0
MS76JL23	2	0.05	0.83	0	1.73	0.23	96.64	2.87	4.61	2.83	1.51	0.0
MS76JL29	2	0.03	1.05	0	1.18	0.10	63.17	1.35	4.61	2.04	0.94	0.0
MS76AG10	2	0.08	0.90	1	2.22	0.34	146.14	1.13	4.17	0.87	2.32	1887.
MS76AG16	2	0.06	0.75	0	1.73	0.58	97.77	0.91	2.92	0.14	1.90	1718.
MS76AG22	2	0.02	0.81	0	0.86	0.27	39.55	1.60	3.86	1.17	1.34	0.0
MS76SP27	2	0.03	0.60	0	1.17	0.89	70.32	1.99	1.69	1.00	0.81	2005.
MS76OC03	2	0.04	0.64	1	1.14	0.68	64.58	0.89	3.77	1.42	2.30	811.
MS76OC21	2	0.02	0.11	0	1.07	0.52	75.46	2.21	5.03	1.09	0.90	1527.
MS76OC27	2	0.04	0.06	0	1.41	0.28	97.96	1.55	2.57	0.50	0.76	1337.
MS76NV02	2	0.06	0.07	0	2.11	1.20	128.89	5.21	3.90	0.84	1.23	3238.
MS76NV08	2	0.07	0.10	0	2.61	1.95	166.91	7.48	4.44	0.25	1.48	6301.
MS76NV14	2	0.02	0.10	0	0.88	1.19	71.35	5.29	4.91	1.80	1.02	2243.
MS76NV20	2	0.02	0.11	0	0.82	1.17	63.51	3.50	4.75	3.48	0.48	3580.
MS76NV26	2	0.06	0.16	0	2.13	1.27	140.33	4.12	5.91	2.11	2.07	3294.
MS76OC08	2	0.10	0.14	0	3.34	2.37	165.80	9.22	5.41	0.97	999999	4630.
MS76OC14	2	0.08	0.20	0	3.60	2.02	157.50	7.62	5.96	0.48	2.24	5871.
MS76OC20	2	0.05	0.17	0	2.03	0.94	102.86	4.29	8.23	1.69	1.17	3150.
MS77JR07	2	0.05	0.08	0	2.38	1.44	105.14	7.17	3.75	1.03	1.80	1623.
MS77FB18	2	0.12	0.23	0	999999	1.76	180.27	6.90	5.71	1.36	1.38	3437.
MS77FB24	2	0.09	0.19	0	999999	0.49	140.75	1.49	3.15	0.0	1.65	2053.
MS77MR06	2	0.08	0.16	0	3.14	1.72	137.76	5.41	3.22	1.04	2.21	4057.
MS77MR14	2	0.10	0.08	0	3.58	1.37	161.03	4.18	2.73	2.15	2.00	3007.
MS77MR20	2	0.06	0.06	0	2.25	0.78	91.89	1.94	2.10	0.07	1.14	2769.
MS77MR26	2	0.03	0.10	0	1.08	0.72	53.12	1.14	1.71	0.0	2.24	3103.
MS77AP01	2	0.03	0.07	0	1.41	0.74	71.72	0.78	2.82	0.73	0.28	3580.
MS77AP07	2	0.05	0.10	0	2.30	0.95	99.35	2.00	7.08	0.47	0.69	3150.
MS77MY01	2	0.05	0.08	0	999999	0.95	76.25	2.58	2.08	1.90	2.96	1432.
MS77MY07	2	0.06	0.07	0	999999	0.50	103.74	2.02	1.67	0.0	6.15	525.
MS77MY13	2	0.06	0.04	0	999999	0.65	92.56	1.52	2.59	0.03	3.67	621.
MS77MY19	2	0.07	0.05	0	999999	0.95	103.81	1.10	1.68	0.95	1.08	859.
MS77JN06	2	0.06	0.09	0	2.60	0.46	133.56	0.89	6.82	0.27	1.70	999999
MS77JN18	2	0.06	0.05	0	2.62	0.67	114.97	1.29	2.06	0.56	3.74	2339.
MS77JN24	2	0.08	0.05	0	3.17	0.55	147.05	0.0	4.56	0.54	2.96	2005.
MS77JL12	2	0.09	0.24	1	2.66	0.31	164.65	1.19	2.79	1.11	5.25	999999
MS77JL30	2	0.07	0.02	0	2.22	0.76	123.49	1.70	4.46	2.46	4.23	1002.
MS77AG05	2	0.10	0.17	0	1.88	1.24	191.70	3.62	3.98	4.23	2.41	2912.
MS77AG11	2	0.07	0.18	0	2.83	0.53	191.99	0.45	7.11	1.56	1.99	1432.
MS77AG17	2	0.03	0.18	0	1.68	0.55	81.21	0.68	3.30	0.0	0.61	1050.
MS77AG23	2	0.06	0.25	0	2.66	0.50	132.46	0.89	4.59	4.13	2.73	1480.
MS77AG29	2	0.12	0.23	0	4.84	1.63	217.10	3.86	2.54	0.03	2.22	2721.
MS77SP04	2	0.05	0.26	0	1.16	0.42	111.49	0.78	9.25	0.0	0.22	668.
MS77SP10	2	0.05	0.17	0	1.88	0.59	96.67	1.20	5.30	0.0	0.21	1241.
MS77SP16	2	0.08	0.20	0	2.96	2.01	153.82	1.64	4.24	0.15	3.37	477.
MS77OC04	2	0.10	0.22	0	4.06	1.85	176.02	7.70	7.83	0.0	1.70	3723.
MS77OC10	2	0.02	0.14	0	1.41	0.51	58.94	1.96	5.39	0.0	0.0	1909.
MS77OC28	2	0.09	0.11	0	8.91	0.97	180.94	2.89	4.78	0.0	3.87	999999
MS77NV03	2	0.10	0.17	0	8.70	2.65	197.64	9.05	10.72	0.18	1.93	999999
MS77NV09	2	0.08	0.14	0	4.43	2.04	166.49	8.59	3.38	0.0	1.62	6110.
MS77NV15	2	0.18	0.19	0	7.62	5.34	270.08	15.52	10.70	0.09	3.30	2005.
MS77NV21	2	0.15	0.12	0	7.49	3.34	259.42	12.56	7.69	0.26	5.97	5633.
MS77NV27	2	0.11	0.13	0	5.03	2.63	166.12	11.61	9.84	5.98	1.71	2769.
MS77OC03	2	0.15	0.22	0	7.06	2.92	241.48	16.80	8.26	0.15	4.03	4201.
NS75JR12	3	0.04	0.05	0	1.18	0.38	45.78	0.36	0.33	0.48	0.13	999999
NS75JR18	3	0.13	0.07	0	3.49	1.64	143.39	3.09	1.96	6.72	1.60	999999
NS75JR24	3	0.15	0.08	0	4.18	1.16	185.50	5.14	3.73	3.65	2.65	999999
NS75JR30	3	0.06	0.07	0	2.13	0.34	90.56	5.24	2.18	0.52	1.24	999999
NS75FB05	3	0.05	0.09	0	1.55	0.93	72.62	1.80	0.88	3.74	0.68	999999
NS75FB11	3	0.06	0.05	0	2.29	0.78	89.79	4.95	1.23	1.86	0.09	999999
NS75FB23	3	0.09	0.07	0	2.98	1.52	109.43	8.05	1.28	3.51	1.24	999999
NS75MR01	3	0.15	0.08	0	4.83	1.79	174.57	7.06	4.89	4.59	2.11	999999
NS75MR07	3	0.13	0.17	0	3.44	1.06	144.68	2.76	5.70	2.87	9.25	999999
NS75MR13	3	0.03	0.16	0	0.65	0.39	42.53	3.71	4.65	4.17	0.80	999999
NS75MR19	3	0.11	0.18	0	2.98	0.93	161.22	3.64	5.86	3.01	2.65	999999
NS75MR25	3	0.68	0.05	1	999999	0.39	294.82	1.29	4.05	7.65	1.85	999999
NS75MR31	3	0.13	0.13	0	4.10	0.38	186.67	2.45	5.73	0.23	1.46	999999

TABLE A.III (*Continued*)

NAME	CAT	MN	CU	CD	FE	PB	PART	ORG	SULF	NIT	CL	CO
NS75AP24	3.	0.12	0.13	0	3.48	0.53	145.01	5.58	3.87	4.46	1.75	999999
NS75AP30	3.	0.12	0.11	0	3.53	0.48	152.29	0.87	0.87	1.59	0.96	999999
NS75MY06	3.	0.10	0.10	0	3.18	0.52	124.70	0.0	0.92	0.0	1.11	999999
NS75MY30	3.	0.04	0.18	0	1.19	0.46	51.02	0.0	0.31	0.57	0.11	999999
NS75JN17	3.	0.92	0.38	0	9999999	0.46	9999999	C.64	9999999	1.76	9.75	999999
NS75JN23	3.	0.16	1.79	0	5.59	0.52	196.02	0.0	2.72	1.78	3.29	999999
NS75JN29	3.	0.03	0.16	0	1.09	0.0	33.46	1.45	0.79	0.74	0.36	999999
NS75JL05	3.	0.08	0.13	0	2.53	0.30	140.40	0.0	4.94	3.66	1.89	999999
NS75JL29	3.	0.20	0.50	0	4.30	0.33	211.47	C.0	4.98	1.58	3.08	999999
NS75AG04	3.	0.15	0.36	0	4.06	1.02	185.73	2.84	9.65	1.99	2.38	999999
NS75AG10	3.	0.17	0.56	0	4.77	0.53	239.82	6.09	8.77	5.48	3.08	999999
NS75AG16	3.	0.09	0.33	0	2.85	0.86	132.10	0.79	7.24	2.95	11.0	999999
NS75AG22	3.	0.18	0.68	0	5.80	0.68	256.79	1.54	6.66	4.55	1.81	999999
NS75AG28	3.	0.24	0.40	0	7.71	0.70	353.59	5.12	5.82	2.40	2.86	999999
NS75SP03	3.	0.14	0.51	0	4.42	0.41	197.21	6.71	12.55	2.69	1.94	999999
NS75SP09	3.	0.08	0.69	0	2.84	1.13	138.53	7.59	10.97	2.11	2.18	999999
NS75SP15	3.	0.10	0.50	0	2.99	0.66	139.74	7.46	5.65	2.69	4.13	999999
NS75SP21	3.	0.12	0.21	0	3.57	0.08	160.75	0.33	5.42	1.53	8.33	999999
NS75SP27	3.	0.20	0.30	0	5.54	1.37	259.13	2.39	12.11	4.20	4.39	999999
NS750C03	3.	0.17	0.31	0	4.96	0.89	227.32	0.28	9.00	0.85	6.84	999999
NS750C09	3.	0.23	0.29	1	6.44	1.04	287.13	6.46	6.83	3.38	5.46	999999
NS750C21	3.	0.07	0.30	0	1.51	0.57	99.56	2.49	4.18	2.09	2.08	999999
NS750C27	3.	0.15	0.24	0	4.29	1.24	183.78	2.38	8.77	1.41	2.60	999999
NS75NV14	3.	0.18	0.23	0	6.52	1.34	274.04	27.23	6.56	0.49	0.49	2291.
NS750C08	3.	0.06	0.12	0	1.42	0.30	69.06	C.0	3.35	1.01	0.32	1384.
NS750C20	3.	0.03	0.0	0	0.76	0.66	63.15	1.19	4.42	0.34	0.11	1623.
NS750C26	3.	0.03	0.0	0	1.22	0.76	55.65	1.19	2.54	0.92	0.49	1480.
NS76JR13	3.	0.10	0.11	0	3.19	1.18	161.36	4.17	5.54	4.43	0.61	1623.
NS76JR19	3.	0.10	0.12	0	3.50	0.93	178.93	2.27	5.92	1.97	1.53	1814.
NS76FB06	3.	0.02	0.26	1	0.30	0.21	18.75	0.0	2.92	1.10	1.32	0.0
NS76FB18	3.	0.08	0.28	0	2.28	0.76	89.95	2.66	3.35	5.25	0.62	1718.
NS76FB24	3.	0.08	0.01	0	2.89	0.57	107.97	0.92	2.36	1.93	0.28	1480.
NS76MR01	3.	0.08	0.22	0	3.20	0.52	124.54	0.69	2.36	2.20	1.19	1527.
NS76MR07	3.	0.03	0.22	0	1.16	0.42	58.08	1.04	4.07	2.42	1.13	3007.
NS76MR13	3.	0.07	0.26	0	2.93	0.38	96.07	C.35	2.03	0.62	0.23	999999
NS76MR19	3.	0.11	0.12	0	3.10	0.76	152.19	1.39	3.02	2.74	1.30	930.
NS76MR25	3.	0.11	0.12	0	3.54	0.68	201.26	0.68	2.79	1.85	2.57	430.
NS76MR31	3.	0.07	0.11	0	2.13	0.43	93.16	2.54	2.82	0.82	1.62	811.
NS76AP12	3.	0.03	0.11	0	1.27	0.0	161.28	2.87	3.07	1.49	3.79	811.
NS76AP24	3.	0.06	0.12	0	1.79	0.56	89.74	0.33	2.10	3.04	1.19	999999
NS76MY06	3.	0.07	0.32	1	2.04	0.26	106.40	0.0	4.01	1.98	1.51	1193.
NS76MY12	3.	0.09	0.25	1	2.14	0.33	108.22	0.0	3.13	2.62	0.70	716.
NS76MY18	3.	0.13	0.27	0	3.34	0.13	177.55	0.0	8.81	2.20	0.27	239.
NS76MY24	3.	0.11	0.26	0	2.54	0.13	115.39	0.55	2.91	0.79	1.89	525.
NS76JN05	3.	0.10	0.14	1	2.80	0.40	115.11	C.0	2.15	1.40	3.44	668.
NS76JN11	3.	0.11	0.11	0	3.46	0.32	153.01	2.87	1.13	1.53	2.70	716.
NS76JN23	3.	0.14	0.13	1	4.23	0.28	157.03	0.88	2.07	1.56	0.43	907.
NS76JL05	3.	0.10	0.09	0	3.89	0.33	164.30	0.22	7.16	1.67	3.61	334.
NS76JL17	3.	0.07	0.05	0	2.43	0.17	105.85	1.43	5.33	3.20	1.58	621.
NS76JL23	3.	0.06	0.06	0	2.04	0.29	99.89	0.11	4.81	4.33	0.74	430.
NS76JL29	3.	0.08	0.08	0	3.32	0.07	151.74	1.54	4.95	2.81	1.03	859.
NS76AG10	3.	0.21	0.12	0	5.82	0.43	294.88	2.54	4.79	1.65	2.05	907.
NS76AG16	3.	0.14	0.29	0	4.08	1.49	128.19	C.0	2.41	0.21	1.24	621.
NS76AG22	3.	0.17	0.05	0	6.10	0.19	266.24	0.65	3.43	1.27	1.80	716.
NS76SP27	3.	0.06	0.10	0	2.38	0.38	141.81	0.0	1.67	1.83	0.47	1227.
NS760C03	3.	0.09	0.10	1	2.94	0.68	129.94	0.0	4.41	1.72	1.90	143.
NS760C21	3.	0.04	0.12	0	1.80	0.31	84.44	C.11	7.99	1.44	0.87	621.
NS760C27	3.	0.05	0.05	0	1.77	0.23	82.01	0.0	1.02	0.54	0.59	525.
NS76NV02	3.	0.06	0.09	0	6.70	0.84	85.98	0.22	3.49	0.17	1.43	1527.
NS76NV08	3.	0.06	0.11	0	2.39	0.48	107.60	0.0	4.41	0.17	0.26	1337.
NS76NV14	3.	0.01	0.15	0	0.44	0.65	34.84	0.54	4.23	1.49	0.79	859.
NS76NV20	3.	0.05	0.15	0	1.65	1.35	102.77	2.66	7.55	5.85	1.02	811.
NS76NV26	3.	0.10	0.20	0	3.35	1.29	161.29	2.21	6.55	3.30	3.49	143.
NS76DC08	3.	0.20	0.11	0	7.91	1.92	247.89	3.78	4.36	3.96	2.65	2083.
NS76DC14	3.	0.13	0.16	0	5.45	1.27	172.36	2.14	6.47	2.01	1.11	1337.
NS76DC20	3.	0.10	0.16	0	3.80	0.77	137.18	1.32	9.97	0.79	1.35	1337.
NS77JR07	3.	0.04	0.15	0	2.12	1.10	75.57	4.80	4.34	0.93	0.0	1146.
NS77FB18	3.	0.18	0.62	0	9999999	1.30	225.11	3.84	5.00	1.36	2.10	999999
NS77FB24	3.	0.21	0.55	0	9999999	0.63	242.13	C.68	2.21	0.18	1.27	999999
NS77MR08	3.	0.12	0.41	0	4.17	1.02	162.18	1.73	4.79	1.14	0.74	999999
NS77MR14	3.	0.12	0.30	0	4.61	0.81	196.73	1.12	3.46	3.73	3.10	999999
NS77MR20	3.	0.15	0.28	0	5.42	0.90	196.75	2.62	3.08	3.34	1.45	999999
NS77MR26	3.	0.03	0.27	0	1.18	0.57	58.16	0.80	1.07	0.06	0.95	477.
NS77AP01	3.	0.09	0.23	0	4.38	0.64	151.14	0.33	3.62	1.53	2.94	1193.
NS77AP07	3.	0.10	0.28	0	3.73	0.95	131.51	3.68	7.32	C.56	0.98	382.
NS77MY01	3.	0.14	0.31	0	9999999	0.42	191.02	2.68	2.31	1.45	3.50	286.
NS77MY07	3.	0.35	0.13	0	9999999	0.56	416.92	1.69	2.47	0.08	7.85	334.
NS77MY13	3.	0.16	0.10	0	9999999	0.73	235.59	1.24	3.04	0.29	4.42	239.
NS77MY19	3.	0.27	0.09	1	9999999	0.92	306.81	5.89	3.67	0.61	3.40	239.
NS77JN06	3.	0.25	0.08	0	9.16	0.33	315.98	0.0	8.96	1.17	2.97	999999

TABLE A.III (*Continued*)

NAME	CAT MN	CU	CD	FE	PB	PART	ORG	SULF	NIT	CL	CO
NS77JN18	3-0.13	0.05	0	4.62	0.48	180.42	0.21	1.47	0.66	2.94	999999
NS77JN24	3-0.28	0.06	0	9.35	0.31	330.96	0.0	6.2'	2.30	3.23	143.
NS77JL12	3-0.38	0.04	1	12.48	0.42	589.31	0.87	4.21	3.68	1.98	999999
NS77JL30	3-0.18	0.06	0	6.57	0.52	245.06	0.0	5.00	3.69	4.26	95.
NS77AG05	3-0.21	0.14	0	7.41	0.82	279.59	0.36	4.05	0.15	0.88	229.
NS77AG11	3-0.16	0.15	0	6.21	0.66	248.92	0.0	4.86	0.68	0.62	521.
NS77AG17	3-0.04	0.19	0	1.92	0.30	88.28	0.0	3.02	1.93	0.23	48.
NS77AG23	3-0.10	0.09	0	3.60	0.56	159.05	0.0	3.78	6.35	1.66	999999
NS77AG29	3-0.15	0.12	0	5.52	0.66	205.12	0.86	2.82	0.39	0.63	999999
NS77SP04	3-0.16	0.15	0	6.19	0.53	224.85	0.11	10.61	0.03	16.5	999999
NS77SP10	3-0.13	0.15	0	5.65	0.35	250.18	0.0	6.27	0.06	3.36	999999
NS77SP16	3-0.05	0.06	0	2.38	0.59	78.90	0.0	1.93	0.0	1.79	999999
NS770C04	3-0.09	0.30	0	8.75	0.78	121.79	0.58	5.49	0.0	0.81	95.
NS770C10	3-0.08	0.25	0	3.32	0.60	121.17	2.15	5.42	0.0	1.01	143.
NS770C28	3-0.10	0.21	0	8.60	0.47	135.76	0.89	4.55	0.08	0.79	191.
NS77NV03	3-0.10	0.24	0	8.97	1.20	158.30	2.95	6.12	0.17	1.10	999999
NS77NV09	3-0.07	0.18	0	3.49	0.85	82.98	2.26	1.97	0.04	0.38	999999
NS77NV15	3-0.17	0.19	0	6.96	1.70	196.52	4.25	5.51	2.07	0.24	999999
NS77NV21	3-0.12	0.15	0	5.92	1.33	155.49	1.47	4.24	0.35	2.69	999999
NS77NV27	3-0.11	0.16	0	5.32	1.21	148.19	2.18	5.07	2.52	2.44	999999
NS77DC03	3-0.15	0.25	0	6.71	1.89	171.27	3.64	3.84	1.21	2.17	999999
SC75JR12	4-0.03	0.91	0	1.08	1.96	63.31	7.77	1.25	0.41	1.68	1623.
SC75JR18	4-0.07	0.77	0	2.06	3.83	112.23	14.23	1.28	5.89	2.73	3103.
SC75JR24	4-0.09	0.88	1	2.65	3.29	136.61	11.37	3.49	1.74	1.68	3866.
SC75JR30	4-0.06	0.74	0	1.80	1.31	91.48	7.71	2.42	1.80	4.37	1814.
SC75FB05	4-0.03	0.62	0	1.58	3.37	88.44	8.91	0.0	3.53	1.17	2503.
SC75FB11	4-0.05	0.42	0	1.70	2.85	89.81	5.74	1.32	0.39	4.61	1146.
SC75FB23	4-0.03	0.57	0	1.30	2.80	78.66	9.40	1.70	2.00	4.00	1671.
SC75MR01	4-0.08	0.33	0	2.28	2.92	122.50	9.11	3.47	2.60	3.29	2434.
SC75MR07	4-0.09	0.33	0	2.25	3.43	136.19	8.92	3.47	2.22	7.22	239.
SC75MR13	4-0.02	0.31	0	0.64	1.49	49.07	4.69	4.37	9999999	2.04	1909.
SC75MR19	4-0.05	0.54	0	1.35	1.78	105.18	6.28	2.10	2.89	1.81	1480.
SC75MR25	4-0.10	0.51	0	1.14	0.70	184.24	4.23	3.25	2.39	1.99	1544.
SC75MR31	4-0.15	1.06	0	4.25	0.61	201.37	4.24	5.26	1.71	2.30	1718.
SC75AP24	4-0.09	1.37	0	2.88	2.42	156.49	5.45	3.44	3.17	3.37	2196.
SC75AP30	4-0.09	0.90	0	2.75	2.49	139.64	5.45	0.0	0.81	2.35	1909.
SC75MY06	4-0.08	1.06	1	2.55	1.83	138.28	1.31	0.34	0.06	0.48	716.
SC75MY30	4-0.07	0.99	0	2.05	1.85	107.72	2.65	0.88	0.85	1.46	2243.
SC75JN17	4-0.13	0.96	0	6.02	0.64	211.68	1.07	3.42	0.55	7.69	381.
SC75JN23	4-0.09	1.20	0	3.71	0.96	141.71	1.60	2.48	1.84	8.16	525.
SC75JN29	4-0.08	1.33	0	2.94	2.05	127.63	5.65	1.70	2.55	5.70	1097.
SC75JL05	4-0.05	1.41	0	1.92	0.83	122.40	0.43	5.97	2.49	4.50	239.
SC75JL29	4-0.12	1.43	0	2.70	1.13	147.21	0.0	4.95	1.72	4.04	143.
SC75AG04	4-0.09	1.24	0	2.95	2.26	163.45	8.15	10.62	3.20	6.25	2100.
SC75AG10	4-0.14	1.33	0	4.07	1.19	233.52	7.02	8.66	5.09	3.89	334.
SC75AG16	4-0.07	1.47	0	2.37	2.30	129.87	3.20	5.27	2.68	5.54	1337.
SC75AG22	4-0.08	1.78	0	2.75	2.36	156.82	3.40	9999999	1.91	4.01	2864.
SC75AG28	4-0.11	2.21	0	3.63	3.03	129.93	3.58	4.38	3.38	4.43	382.
SC75SP03	4-0.06	1.34	0	1.58	0.89	88.61	1.16	7.89	1.40	3.09	644.
SC75SP09	4-0.03	0.44	0	0.68	1.14	64.87	6.82	6.21	2.12	2.45	621.
SC75SP15	4-0.04	0.09	0	1.38	1.74	72.51	7.89	3.73	0.94	3.73	573.
SC75SP21	4-0.04	0.08	0	1.14	0.62	65.65	1.76	4.16	1.09	1.27	0.0
SC75SP27	4-0.09	0.13	0	2.94	2.73	134.44	4.21	6.22	2.50	3.05	1050.
SC750C03	4-0.09	0.21	0	2.46	2.71	142.94	4.40	10.43	0.82	5.89	2482.
SC750C09	4-0.11	0.16	1	3.18	1.96	165.37	6.78	4.97	1.95	3.27	2005.
SC750C21	4-0.14	0.14	0	4.14	3.44	212.81	8.94	5.94	0.74	5.26	1432.
SC750C27	4-0.09	0.19	0	2.92	2.60	155.97	0.44	6.27	2.96	3.30	1002.
SC75NV14	4-0.13	0.15	0	4.24	5.60	281.07	42.75	2.04	0.78	4.96	7671.
SC75DC08	4-0.08	0.11	0	1.78	3.46	127.10	12.21	4.35	4.62	3.00	2673.
SC75DC20	4-0.02	0.02	0	0.88	1.61	71.75	6.64	5.68	0.54	1.75	1575.
SC75DC26	4-0.04	0.0	0	1.62	3.78	122.47	22.37	2.07	1.52	2.82	3485.
SC76JR13	4-0.07	0.15	0	2.24	3.37	162.95	18.04	5.50	4.80	999999	2912.
SC76JR19	4-0.06	0.09	0	1.99	1.92	116.17	5.26	4.38	1.29	1.63	1623.
SC76FB06	4-0.07	0.13	0	2.24	4.13	187.89	14.13	12.05	3.83	1.08	764.
SC76FB18	4-0.06	0.33	0	2.20	3.26	97.37	5.71	3.65	6.20	2.36	1213.
SC76FB24	4-0.06	0.36	0	2.58	2.94	111.21	6.30	1.97	1.69	2.07	3389.
SC76MR01	4-0.05	0.25	0	2.18	1.20	120.79	3.57	1.34	2.08	1.05	1957.
SC76MR07	4-0.02	0.17	0	1.14	1.37	75.32	4.91	2.99	2.23	1.49	716.
SC76MR13	4-0.04	0.22	0	1.62	1.98	85.59	5.18	2.79	0.46	3.15	797.
SC76MR19	4-0.05	0.12	0	1.78	1.45	100.27	3.30	1.85	1.88	2.23	1384.
SC76MR25	4-0.07	0.14	0	2.05	1.39	135.34	4.46	1.70	1.73	3.15	621.
SC76MR31	4-0.07	0.23	0	1.84	1.91	114.18	4.26	3.41	0.87	3.12	1909.
SC76AP12	4-0.07	0.24	0	2.33	0.83	130.93	3.58	2.85	2.46	6.27	907.
SC76AP24	4-0.06	0.50	0	2.05	2.12	132.57	6.44	2.93	4.07	1.80	1575.
SC76MY06	4-0.08	0.58	1	1.73	0.84	112.21	1.70	3.44	2.43	4.22	668.
SC76MY12	4-0.06	0.55	1	1.47	0.99	97.86	1.78	3.09	2.59	1.92	811.
SC76MY18	4-0.05	0.48	0	1.44	0.54	92.64	1.16	8.09	2.10	1.01	430.
SC76MY24	4-0.08	0.51	0	1.73	0.80	105.44	1.78	2.54	1.14	1.62	525.
SC76JN05	4-0.05	0.54	0	1.45	1.12	96.07	4.19	1.75	2.04	3.28	1050.
SC76JN11	4-0.08	0.39	0	2.64	1.32	150.60	5.32	1.80	1.08	4.00	525.

TABLE A.III (*Continued*)

NAME	CAT	MIN	CU	CD	FE	PB	PART	ORG	SULF	NIT	CL	CO
SC76JN23	4	0.08	0.35	0	2.12	1.46	132.35	4.96	2.98	1.91	1.93	1337.
SC76JL05	4	0.11	0.28	0	3.76	0.71	171.50	2.36	8.55	1.33	2.41	99999
SC76JL17	4	0.05	0.22	0	1.79	0.50	104.58	2.52	6.43	3.26	1.93	0.0
SC76JL23	4	0.09	0.69	0	2.87	1.23	192.94	6.92	17.58	3.24	0.69	0.0
SC76JL29	4	0.05	0.25	0	1.93	0.17	103.32	1.83	4.92	3.00	1.37	99999
SC76AG10	4	0.13	0.24	0	3.58	0.88	198.96	2.45	6.25	1.52	3.01	1483.
SC76AG16	4	0.02	0.02	0	0.77	0.0	190.09	2.79	3.20	1.00	2.64	2578.
SC76AG22	4	0.07	0.21	0	2.46	0.49	124.44	2.57	5.36	1.39	2.71	668.
SC76SP27	4	0.01	0.03	0	0.25	0.03	20.41	0.0	0.59	0.42	0.0	2649.
SC76OC03	4	0.05	0.10	1	2.03	1.87	111.76	4.22	2.12	1.58	1.30	1814.
SC76OC21	4	0.02	0.17	0	1.37	1.86	85.13	3.17	8.23	1.57	1.70	2912.
SC76OC27	4	0.03	0.08	0	1.14	0.62	72.98	1.26	2.37	0.75	0.57	239.
SC76NV02	4	0.02	0.08	0	0.79	1.35	50.51	2.45	2.54	0.68	2.12	3819.
SC76NV08	4	0.08	0.15	0	2.66	2.97	156.72	4.96	6.65	0.48	3.45	3437.
SC76NV14	4	0.03	0.23	0	0.90	2.17	84.90	10.00	3.52	2.12	1.54	3198.
SC76NV20	4	0.07	0.25	0	2.63	5.53	179.92	16.12	10.07	6.72	4.03	3780.
SC76NV26	4	0.10	0.12	0	3.27	1.11	197.10	4.81	6.09	1.95	4.12	4888.
SC76DC08	4	0.13	0.21	0	4.51	6.16	237.35	19.94	5.06	2.83	9.23	5967.
SC76DC14	4	0.10	0.25	0	4.16	4.73	189.79	13.29	8.56	0.71	6.42	4296.
SC76DC20	4	0.04	0.16	0	1.92	2.12	101.75	6.11	10.33	1.47	2.55	1098.
SC77JR07	4	0.04	0.12	0	1.98	4.12	112.54	16.06	4.96	1.22	4.57	4105.
SC77FB18	4	0.11	0.35	0	99999999	3.66	224.78	10.50	5.63	0.61	4.52	4296.
SC77FB24	4	0.08	0.22	0	99999999	1.19	132.26	2.43	2.18	0.21	3.30	1957.
SC77MR08	4	0.12	0.24	0	4.61	3.08	193.31	8.28	4.43	0.75	5.10	4057.
SC77MR14	4	0.09	0.13	0	3.39	2.58	140.77	6.14	2.52	2.85	4.03	764.
SC77MR20	4	0.08	0.15	0	3.24	1.72	137.58	3.83	1.84	0.0	1.83	2291.
SC77MR26	4	0.03	0.12	0	1.12	0.99	57.27	2.79	1.42	0.47	1.65	2053.
SC77AP01	4	0.05	0.09	0	1.84	0.93	89.33	2.00	3.19	1.61	1.54	1289.
SC77AP07	4	0.07	0.18	0	2.99	0.20	117.74	4.96	7.84	1.64	1.95	2434.
SC77MY01	4	0.04	0.09	0	99999999	1.20	88.50	3.57	1.86	1.51	5.06	1384.
SC77MY07	4	0.06	0.08	0	99999999	1.00	102.33	0.50	1.94	0.09	10.1	1050.
SC77MY13	4	0.05	0.11	0	99999999	1.04	69.19	2.26	2.34	0.57	6.64	1480.
SC77MY19	4	0.07	0.13	1	99999999	1.44	108.86	2.02	3.02	0.57	4.07	859.
SC77JN06	4	0.05	0.12	0	1.71	0.77	109.71	0.90	12.05	1.14	4.91	1623.
SC77JN18	4	0.07	0.06	0	3.00	1.18	141.20	2.38	2.90	2.61	5.49	1098.
SC77JN24	4	0.09	0.09	0	3.32	0.88	155.73	1.78	4.28	1.06	3.74	1718.
SC77JL12	4	0.09	0.07	1	2.70	0.66	154.09	1.43	4.04	1.82	4.51	99999
SC77JL30	4	0.05	0.05	0	1.82	0.62	107.45	1.12	4.43	2.44	5.66	573.
SC77AG05	4	0.07	0.05	0	1.43	1.87	134.14	4.01	3.81	2.83	4.35	1575.
SC77AG11	4	0.05	0.06	0	1.85	0.69	103.31	0.34	4.72	0.42	1.86	286.
SC77AG17	4	0.02	0.05	0	0.99	0.52	49.46	0.0	2.40	0.99	1.27	668.
SC77AG23	4	0.04	0.06	0	1.31	0.69	79.15	1.04	3.06	4.42	1.87	525.
SC77AG29	4	0.08	0.08	0	3.01	1.72	132.32	3.38	3.00	0.02	2.99	1098.
SC77SP04	4	0.06	0.08	0	1.11	1.13	99.66	1.52	10.55	0.0	1.91	1193.
SC77SP10	4	0.03	0.05	0	1.22	0.63	62.99	0.76	5.07	0.0	1.41	143.
SC77SP16	4	0.07	0.07	0	2.29	2.31	109.69	1.18	3.46	2.38	2.74	143.
SC77OC04	4	0.05	0.07	0	5.23	1.48	96.79	2.71	5.46	0.0	0.0	2100.
SC77OC10	4	0.03	0.05	0	1.29	1.94	72.76	5.67	4.52	0.0	0.0	1337.
SC77OC28	4	0.06	0.07	0	5.59	1.49	107.70	2.50	4.44	0.06	2.20	1718.
SC77NV03	4	0.08	0.11	0	7.35	4.01	153.54	7.82	7.82	0.08	2.57	3962.
SC77NV09	4	0.06	0.05	0	3.08	2.66	106.81	7.10	2.46	0.02	2.55	3007.
SC77NV15	4	0.15	0.14	0	5.58	5.70	198.11	10.91	7.20	0.68	2.59	6349.
SC77NV21	4	0.11	0.08	0	4.75	4.97	175.00	11.40	5.41	0.29	6.08	5251.
SC77NV27	4	0.10	0.09	0	4.11	3.94	146.22	13.34	8.77	6.17	2.67	4344.
SC77OC03	4	0.07	0.07	0	3.66	1.98	108.72	5.37	3.72	0.11	3.95	4582.

ƎFIN

APPENDIX IV
INDICES OF BMDP, SPSS, AND
ARTHUR PACKAGES

BMDP PROGRAMS

P1D:	Simple data description
P2D:	Detailed data description including frequencies
P3D:	Comparison of groups with *t*-tests
P4D:	Single column frequencies
P5D:	Histograms and univariate plots
P6D:	Bivariate plots
P7D:	Description of groups (histograms and analyses of variance)
P8D:	Missing values
P9D:	Multiway group description
P1F:	Two-way frequency tables
*P2F:	Two-way frequency tables
*P3F:	Multiway frequency tables
*P1L:	Life tables and survival functions
P1M:	Cluster of variables
P2M:	Cluster of cases
P3M:	Block clustering
P4M:	Factor analysis
P6M:	Canonical correlations
P7M:	Stepwise discriminant analyses
*PAM:	Estimations for missing data
†PKM:	*K*-means clustering
P1R:	Multiple linear regression
P2R:	Stepwise regression
P3R:	Nonlinear regression
P4R:	Principal components
P5R:	Polynomial regression

P6R: Partial correlations
*P9R: All possible subsets regression
*PAR: Derivative-free nonlinear regression
†PLR: Stepwise logistic regression
P1S: Multipass transformation
P2S: Nonparametric statistics
P1V: One-way analysis of variance and covariance
P2V: Variance and covariance (with repeated measures)
*P3V: Mixed model analysis of variance
†P8V: General mixed model analysis of variance

*Added in the 1977 P-series.
†Added in the 1979 Manual for the P-series.

SPSS PROGRAMS (1975 SECOND EDITION)

AGGREGATION: Tabular output of aggregated characteristics
ANOVA: Multiple-way analysis of variance
BREAKDOWN: Intervariable relations with table displays
CANCORR: Canonical correlations
CONDESCRIPTIVE: Central tendency and dispersion measurements
CROSSTABS: Cross-tabulations of variables
DISCRIMINANT: Discriminant analysis
FACTOR: Factor analysis
FREQUENCIES: Descriptive statistics for noncontinuous data
GUTTMAN SCALE: Special relationships between variables
NONPAR CORR: Nonparametric correlations
*NPAR TESTS: Nonparametric tests
ONEWAY: One-way analysis of variance
PARTIAL CORR: Partial correlations
PEARSON CORR: Zero-order product–moment correlation coefficients
REGRESSION: Multiple regression analysis
SCATTERGRAM: Two-dimensional plots
T-TEST: T-test sample means comparisons

*Added in Releases 7 and 8

ARTHUR PROGRAMS

BAYES: Bayesian statistical calculations
CORREL: Pearson product–moment correlation coefficients
CHANGE: For changes in data structure

DISTANCE: Interpattern distance calculations
GRAB: Grabbing important variables
HIER: Hierarchical clustering technique
KARLOV: Karhunen–Loeve transformation
KNN: K-nearest-neighbors method
LEAST: Least-squares calculations
MULTI: Multiplane separation
NLM: Nonlinear mapping
PIECE: Piecewise linear regression
PLANE: Two-category planar separations
PNN: Percent nearest neighbors
SCALE: Z-score transformation scaling
SELECT: Variable selection
SIMCA: Statistical isolinear multiple components analysis
STEP: Stepwise regression
TREE: Minimal spanning tree
VARVAR: Two-dimensional variable plot
WEIGHT: Variable weighting calculations

APPENDIX V
PROGRAMS, MANUALS, AND
REFERENCE INFORMATION

A. PROGRAMS AND SUPPORT

1. SPSS, Inc.
   Suite 3300
   440 N. Michigan Avenue
   Chicago, Illinois 60611

2. BMDP
   Health Science Computing Facility
   AV-111, CHS
   University of California
   Los Angeles, California 90024

3. ARTHUR
   Infometrix, Inc.
   P.O. Box 25808
   Seattle, Washington 98125

4. CLUSTAN
   16 Kingsburgh Road
   Edinburgh EH12 6DZ
   Scotland

5. SAS
   SAS Institute Inc.
   Box 8000
   Cary, North Carolina 27511

B. MANUALS

1. *SPSS, Statistical Package for the Social Sciences, Second Edition*
Norman H. Nie, C. H. Hull, J. G. Jenkins, K. Steinbrenner, and D. H. Bent
ISBN 0-07-046531-2
Copyright 1975 McGraw-Hill, Inc.

   SPSS Statistical Algorithms, Release 8.0
M. J. Norusis
Copyright 1979 by SPSS, Inc.

   SPSS Update: New Procedures and Facilities for Releases 7 and 8
C. Hadlai Hull and Norman H. Nie
ISBN 0-07-046534-7
Copyright 1979 by Hull and Nie
Available from McGraw-Hill, Inc.

2. *BMDP Biomedical Computer Programs P-Series 1977*
W. J. Dixon (Series Editor), M. B. Brown (1977 Edition)
ISBN 520-03569-0
Copyright 1977 by the Regents of the University of California
Available from University of California Press,
2223 Fulton Street, Berkeley, California 94720
The 1979 printing of the above book contains three additional programs.

3. *Documentation for ARTHUR*
Distributed by Infometrix, Inc.

4. *CLUSTAN User's Manual*
D. Wishart, Inter-University Research Councils Series
Report No. 47, January, 1978
ISBN 0-906296-00-5
Available from: Program Library Unit, Edinburgh University,
18 Buccleuch Place, Edinburgh E H8 9LN Scotland

5. *SAS User's Guide, 1979 Edition*
William H. Blair
ISBN 0-917382-06-3
Copyright 1979 by SAS Institute, Inc.
Available from Publications Department of SAS Institute, Inc.

   SAS Introductory Guide
Also available: *SAS Programmer's Guide, 1979 Edition*
   SAS Views
   SAS Users Group Conference Proceedings
   SAS Applications Guide
   SAS/GRAPH and *SAS/ETS* User's Guide

C. ADDITIONAL REFERENCES

1. *SPSS Pocket Guide, Release 8*
   Available from address in A.

   Keywords (the SPSS newsletter)
   Available from the address in A.

2. *BMDP User's Digest* (a condensed guide to the programs)
   Prepared by Mary Ann Hill
   Available from BMDP Statistical Software, Department of
   Biomathematics, UCLA, Los Angeles, California 90024

   BMDP Technical Reports
   A list is available from the address in A.

   BMDP Statistical Software Communications
   A noncost publication series available from address in A.

3. *Chemometrics Newsletter (a noncost publication)*
   Sergio Clementi
   Instituto di Chimica Organica
      dell' Universitá di Perugia
   Via Eke di Sotto 10
   Perugia, Italy

4. *CLUSTAN Technical Specification (a condensed guide)*
   Available from address in A.

5. *SAS Technical Report Series*
   A list is available from the Publications Department
   of SAS Institute, Inc.

   SAS Reference Card (a condensed summary)
   Available from SAS Institute, Inc.

   SAS Communications (the SAS newsletter)
   Available from the address in A.

APPENDIX VI
SUMMARY OF ANALYSES AND
PROGRAM CROSS-REFERENCE
FOR CHAPTER II

A summary of the approach taken for data analysis will be given. The first program names refer to those most commonly used. Those enclosed in double parentheses are comparable ones available in the other packages. This choice was made taking into consideration ease of input and output interpretation and statistical choices available in the programs. A condensed summary is given at the end of the appendix followed by a cross-reference for BMDP, ARTHUR, and SPSS.

II.1. INTRODUCTORY DATA EXAMINATIONS

Goal: Data screening and distributional characteristics of each variable.

II.1a: BMDP4D—Simple first data screening for errors.

II.1b: CONDESCRIPTIVE (SPSS)—To check format compatibility and further data screening.

II.1c: BMDP1D—Data listing.

II.1d: BMDP2D—Study Gaussian assumptions, typical values, and unique points.

II.1e: BMDP1D—Identify missing, minimum, and maximum values during data listing.

II.1f: BMDP2D—To recheck the Gaussian assumptions after removing atypical points.

II.1g: CONDESCRIPTIVE (SPSS)—To find distributional characteristics of the data.

((SCALE-ARTHUR, FREQUENCIES-SPSS, BMDP5D))

II.2. STRATIFICATION OF VARIABLES

Goal: To find out if the data as a whole are stratified into groups and check the statistical differences between the groups.

II.2a: BMDP3D—Compare pairs of classes and their means to find univariate statistics and histograms within each group.
((TTEST-SPSS))

II.2b: BMDP7D—Compare single variable in any number of groups simultaneously for equality of means and variances.
((ANOVA and ONEWAY-SPSS))

II.2c: BMDP9D—To study the breakdown of a given variable between defined groups of another variable (or variables)
((BREAKDOWN-SPSS))

Optional: BMDP1F, BMDP2F, BMDP3F—Frequency tables ((CROSSTABS-SPSS))

II.3. INTERVARIABLE RELATIONSHIPS

Goal: To study pairs and sets of variables for relationships between them.

II.3a: PEARSON CORR (SPSS)—To study all bivariate correlations in the data.
((CORREL-ARTHUR, BMDP6D))

II.3b: SCATTERGRAM (SPSS)—To create bivariate plots of the data and obtain simple regression lines.
((VARVAR-ARTHUR, BMDP6D))

II.3c: PARTIAL CORR (SPSS)—To study bivariate correlations while controlling for other variables.
((BMDP6R))

II.3d: BMDP1M—Clustering of variables are studied and dendrograms plotted.

II.3e: BMDP6M—To study relationships between two groups of variables.
((CANCORR-SPSS))

II.3f: BMDP2R or REGRESSION (SPSS)—To predict values of one variable with linear combinations of the remaining ones.
((BMDP1R, BMDP9R; nonlinear—BMDP3R, BMDP5R, BMDPAR))

II.4. UNSUPERVISED LEARNING

Goal: To study natural groupings present in the data considering all variables simultaneously.

II.4a: TREE (ARTHUR)—To draw a tree diagram to find natural groupings of patterns.

II.4b: HIER (ARTHUR)—To use hierarchical clustering techniques for finding natural groupings.

((BMDP2M, BMDPKM))

II.4c: NLM (ARTHUR)—To plot *n*-dimensional space into two dimensions preserving interpattern distances.

II.4d: CLUSTAN—Introduction of a fourth program dedicated solely to unsupervised learning techniques.

II.5. SUPERVISED LEARNING

Goal: To train the computer to recognize groups in categorized data and determine separability.

II.5a: KNN (ARTHUR)—Uses a "nearest-neighbor" technique to define group membership for the patterns.

II.5b: DISCRIMINANT (SPSS)—Uses a linear combination of variables to define a surface that best separates patterns into defined categories.

((BMDP7M, MULTI-ARTHUR, PLANE-ARTHUR))

II.6. DATA REDUCTION

Goal: To reduce the data into a smaller number of meaningful variables.

II.6a: SELECT/VARVAR (ARTHUR)—To reduce the number of individual variables.

II.6b: BMDP4R—To reduce the number of variables by creating a smaller number of linear combinations (no underlying mathematical models required).

((KARLOV-ARTHUR, FACTOR-SPSS))

II.6c: FACTOR (SPSS)—To form linear combinations of variables to discover underlying causes for the data patterns.

((BMDP4M))

TABLE A.VI.a
Summary of Comparable Procedures for Each Step[a]

	BMDP	SPSS	ARTHUR
II.1a	4D		
II.1b		CONDESCRIPTIVE	
II.1c	1D		
II.1d	2D		
II.1e	1D		
II.1f	2D		
II.1g	(5D)	CONDESCRIPTIVE (FREQUENCIES)	(SCALE)
II.2a	3D	(TTEST)	
II.2b	7D	(ANOVA, ONEWAY)	
II.2c	9D	(BREAKDOWN)	
Optional	1F,2F,3F	(CROSSTABS)	
II.3a	(6D)	PEARSON CORR	(CORREL)
II.3b	(6D)	SCATTERGRAM	(VARVAR)
II.3c	(6R)	PARTIAL CORR	
II.3d	1M		
II.3e	6M	(CANCORR)	
II.3f	2R	REGRESSION	
	(1R,9R,3R,5R,AR)		
II.4a			TREE
II.4b	(2M,KM)		HIER
II.4c			NLM
II.4d	CLUSTAN was used at this step		
II.5a			KNN
II.5b	(7M)	DISCRIMINANT	(MULTI, PLANE)
II.6a			SELECT/VARVAR
II.6b	4R	(FACTOR)	KARLOV
II.6c	(4M)	FACTOR	

[a]NOTE: Parentheses indicate those programs which were used less frequently in the steps.

TABLE A.VI.b
Alphabetical Cross-Reference

BMDP		SPSS		ARTHUR	
Program	Step	Program	Step	Program	Step
1D	II.1c,e	ANOVA	II.2b	CHANGE	II.7
2D	II.1d,f	BREAKDOWN	II.2c	CORREL	II.3a
3D	II.2a				
4D	II.1a	CANCORR	II.3e	HIER	II.4b
5D	II.1g	CONDESCRIPTIVE	II.1b,g	KARLOV	II.6b
6D	II.3a,b	CROSSTABS	II opt.	KNN	II.5a
7D	II.2b				
8D	Appx. VIII	DISCRIMINANT	II.5b	MULTI	II.5b
9D	II.2c	FACTOR	II.6c	NLM	II.4c
1M	II.3d	FREQUENCIES	II.1g	PLANE	II.5b
2M	II.4b				
4M	II.6c	NONPAR CORR	Appx. VII	SCALE	II.1g
6M	II.3e	ONEWAY	II.2b	SELECT	II.6a
7M	II.5b	PARTIAL CORR	II.3c	TREE	II.4a
AM	Appx. VIII				
KM	II.4b	PEARSON CORR	II.3a	VARVAR	II.3b
1R	II.3f	REGRESSION	II.3f	WEIGHT	Appx. IX
2R	II.3f	SCATTERGRAM	II.3b		
3R	II.3f	TTEST	II.2a		
4R	II.6b				
5R	II.3f				
6R	II.3c				
9R	II.3f				
AR	II.3f				
LR	II.3f				
1F	II opt.				
2F	II opt.				
3F	II opt.				
3S	Appx. VII				

APPENDIX VII
NONPARAMETRIC STATISTICS

Data are usually assumed to be parametric. This means that the assumption of a Gaussian distribution holds (see Appendix II) and therefore Gaussian statistics such as means, standard deviations, Pearson product–moment correlations, and t-tests can be correctly applied to the data. The assumption is also made that the distribution is mono-modal, which means that only a single central value is present in the data. Such assumptions are easily checked with BMDP4D (see Step II.1d). If these are not correct, but parametric methods are still utilized, the results can be incorrect and lead to false conclusions.

There are distribution-free nonparametric methods of statistical analyses available for such data sets. If, after Steps II.1d through II.1f are applied, plots still are distributed in a non-Gaussian form, the programs discussed here should be utilized in place of Steps II.2 and II.3a. An equivalent form of the parametric programs used in Steps II.2 and II.3a using rank-order nonparametric statistics can be found in either SPSS NONPAR CORR or BMDP3S. ARTHUR contains no programs for such calculations. Releases 7 and 8 of SPSS have added a new program, NPAR TESTS, which has available 14 different nonparametric tests. These will be discussed later.

BMDP3S contains the following:

a. The sign test and Wilcoxon sign-rank test (equivalent to paired t-tests in parametric statistics);
b. Mann–Whitney rank sum tests (equivalent to two-sample t-tests);
c. Kruskal–Wallis one-way analysis of variances;
d. Friedman's two-way analysis of variances;
e. Spearman and Kendall rank-order correlation coefficients.

Only Spearman and Kendall rank-order correlation coefficients are available in NONPAR CORR in SPSS. In NPAR TESTS from Releases 7 and 8, all of the above approaches are available with nine additional ones present, including the Kolmogorov–Smirnov test. Details of the methods are given in the BMDP manual (for all of the above techniques) or the SPSS manual for the correlation statistics.

To apply these methods, values need only have a "ranking" assigned to them so they can be arranged in increasing order. The rankings are then compared instead of the actual values as in parametric statistics.

Assume that the values for a given variable are measured on 15 cases. The results are as follows:

Case No.:	1	2	3	4	5	6	7	8	9	10	11	12	13	14	15
Value:	1	2	6	3	7	2	3	1	7	4	2	4	2	1	3

Nonparametric statistics will then rank these as follows

Rank	Value	Cases
1	1	1,8,14
2	2	2,6,11,13
3	3	4,7,15
4	4	10,12
5	6	3
6	7	5,9

Rank values now range from one to six. These are used for all further calculations. Note that this distribution does not follow a Gaussian curve (or even closely approximate it). Therefore, applications of statistical tools based on Gaussian assumptions could lead to incorrect results. Nonparametric statistics would have to be applied. Repeated rankings (i.e., the value of "1" occurs in cases #1, 8, and 14) can be present in the data. Corrections are made for these types of "ties."

Table A.VII gives results for both Spearman and Kendall nonparametric rank-order correlation coefficients for the air pollution data base from BMDP1S. Results can be compared to those obtained from the parametric Pearson product–moment correlation coefficients obtained in Step II.3a (see Table II.3a). Note that for this example Spearman values are closer to Pearson product values and Kendall values tend to be consistently smaller. MN, FE, and PART are highly correlated with all three methods as are PB, ORG, and CO. It has been found that the Spearman values yield closer approximations to the values from Step II.3a as the data approach a Gaussian distribution, as is the case in this example. As more repeated rankings and non-Gaussian distributions are present, the Kendall values become more meaningful.

Results from NONPARR CORR in SPSS are similar. It is suggested that nonparametric tests such as these and the additional ones in NPAR TESTS be considered. Comparisons can be made to the parametric results for these from Steps II.2.

The choice between parametric and nonparametric statistics should be carefully considered. Erroneous answers are often obtained if the correct usage is not made, resulting in data misinterpretation. This problem in chemistry is addressed by Ames and Szony. General reference books for nonparametric statistics are also available.

TABLE A.VII

Nonparametric Data Analysis from BMDP3S

SPEARMAN RANK CORRELATION COEFFICIENTS

		MN	CU	FE	PB	PART	ORG	SULF	NIT	CL	CO
MN	2	1.0000									
CU	3	.0776	1.0000								
FE	4	.8652	-.0677	1.0000							
PB	5	.2732	.1784	.2968	1.0000						
PART	6	.8700	.0714	.8263	.3510	1.0000					
ORG	7	.1960	.2909	.1839	.7983	.3230	1.0000				
SULF	8	.2856	.0496	.3143	.2006	.3542	.1969	1.0000			
NIT	9	.1021	.2561	-.0261	.0450	.1283	.1285	.1026	1.0000		
CL	10	.3780	.1891	.3348	.4074	.4428	.3533	.2251	.2295	1.0000	
CO	11	.1176	.0916	.1770	.6148	.2350	.6127	.0906	-.0947	.2310	1.0000

KENDALL RANK CORRELATION COEFFICIENTS

		MN	CU	FE	PB	PART	ORG	SULF	NIT	CL	CO
MN	2	1.0000									
CU	3	.0547	1.0000								
FE	4	.7268	-.0454	1.0000							
PB	5	.1969	.1228	.2030	1.0000						
PART	6	.7362	.0480	.6646	.2471	1.0000					
ORG	7	.1438	.1995	.1313	.6148	.2291	1.0000				
SULF	8	.1967	.0354	.2088	.1331	.2340	.1302	1.0000			
NIT	9	.0717	.1764	-.0125	.0322	.0884	.0877	.0732	1.0000		
CL	10	.2709	.1278	.2284	.2842	.3063	.2537	.1527	.1573	1.0000	
CO	11	.0836	.0602	.1189	.4522	.1630	.4425	.0573	-.0629	.1577	1.0000

APPENDIX VIII
MISSING VALUES

In the ideal data base, all of the variables to be considered are measured on each pattern. These variables are then related to some property or grouping in the data. The statistical study proceeds to find these relationships. In the normal case, values for some variables are not available for every pattern. These, then, are considered as missing data and some method must be used to handle these cases.

In ARTHUR, missing values are usually coded in the data cards by using a string of 9's across the entire section of format reserved for the missing value for that given case. This clues the computer to recognize this as a missing value. ARTHUR then uses a simple mean value which that variable takes on for the entire data set. This is used in all subsequent calculations.

There are advantages and disadvantages to this procedure. If the case is not to be deleted from the data base for having missing values, some type of value must be substituted for it so that the necessary mathematical manipulations may be completed. A simple mean for that variable over the entire data set is usually rather safe. The mean is a reasonable guess for the value, and as long as too many variables within a given case or too many measurements of a single variable are not missing from the data base, no real problems are usually encountered. However, an average of the variable within a given group (if the categories are known *a priori*) may be a better value to use.

In the 1-9-77 ARTHUR version, new methods for handling missing values have been added. If "INFILL" is chosen, the mean for the category of which the data piece is a member is used to fill in the missing data. This is usually a better estimate than the mean for the total data base. "INPICK" fills missing data by random selection from the values that feature takes on for the category of the data vector. The "INPCFI" choice allows principal component models to be used for filling the data in question. "INSQGA" assumes a Gaussian distribution for the category about the category mean, and chooses a random value from this distribution to replace this missing value. It is felt that the most useful choice of these would be the mean value for that variable within its category.

In SPSS, one of the control cards for defining the problem is used to define missing values. In this case, the designation for missing values for each variable need not be the same. On the control card, the user may specify three values for each variable in the file that are used to indicate missing data. In this method, the three different accepted missing values may be used to designate to the user why the data were not available. (In Releases 7 and 8 of SPSS, the specification "THRU" is acceptable, and specifications such as "LOWEST THRU 0" are acceptable and count as two of the specifications.) When missing values are encountered by the computer in this case, a choice of ways to treat the data can be made, and are often designated in the options list of SPSS (see Section III.2). Usually a choice of listwise or pairwise deletion is available. In listwise deletion, the entire case is omitted from the calculation if any of the variables listed for input are missing. The program usually will state how many of the original cases were omitted. In pairwise deletions, only those cases which have missing values for those variables needed for the immediate calculation are omitted. A third option (usually not chosen) is to ignore the missing value flags and consider them as normal data. This could lead to misinterpretation of the results.

To illustrate these options, assume the data base consists of three variables and simple correlation coefficients are to be calculated. Assume that the variables are questions from a doctor's survey of his patients. Fifty patients are interviewed. Assume that for question 1, all 50 patients answered. For question 2, 40 answered one of his choices, five did not answer the question at all, and five filled in their own answers. On question 3, 44 answered with one of the doctor's choices and six indicated that it did not apply to their situation.

With ARTHUR, all 50 cases would be considered for all statistical analyses. For question 2, the values for the 40 answers on the questionnaires would be averaged and used as the value for question 2 on the other ten. On question 3, the six questionnaires which indicated that the question did not apply to their situation were still assigned the value of the simple average for the other 44 questionnaires (as if that had been their answer all along). ARTHUR shows no distinction between these types of missing data.

In SPSS, flags can be used for different reasons for not having certain data. If a 0 is used for "no answer," a 1 for "other answer," and a −1 for "does not apply," the data can be coded into the computer in this form. The computer is then told that a 0 or 1 for question 2 indicated missing data and a −1 for question 3. Up to three flags can be used for each variable. This reminds the user, when looking at the data, why it was missing. Assume no deletion was chosen in SPSS. This would result in false answers, since −1, 0, and 1 would then be manipulated as real numbers and not missing value flags. If listwise deletion was chosen, all cases with any of the three values flagged as missing would be deleted, and the analyses would be done on the remainder of the data. If pairwise deletion was chosen, for the correlation of 1 versus 2, 40 cases would be used; for 1 versus 3, 44 cases; and for 2 versus 3, only the cases where neither 2 nor 3 was flagged as missing would be considered. This last choice has the advantage of using as much data as possible for the calculation. However, in comparing the three correlation coefficients it must be remembered that they were not calculated on the same number of cases. Again, if the amount of missing data is minimal, both of the last two methods should give similar results.

BMDP has two programs for handling missing data, BMDP8D and BMDPAM. A very informative technical report on these two programs is available from BMDP.* Both values that are truly missing and those that are outside a stated range (which may be due to errors in the analysis) can be considered as missing. Extreme care must be used in defining this range if this option is chosen.

In BMDP, codes for missing values are given in the VARIABLE paragraph (see Section III.3). Only one code per variable may be used. These missing values can be manipulated in a variety of ways.

With BMDP8D the correlation matrix is calculated and used in subsequent programs. Choices are available for handling the missing data. In one choice, a listwise deletion is made and only cases with a complete analysis list are considered. In the second method, pairwise deletion is used to calculate either correlations or covariances. In the third method, similar to ARTHUR, all acceptable values are used to calculate means and variances. Covariances are then calculated using deviations from these means. A new program in the BMDP series, BMDPAM, seems very useful for further types of studies on missing data. BMDPAM studies the extent to which the data are missing and checks to see if this occurs randomly among the cases, or if some pattern among the cases having missing data exists. (For instance, was it all the males who did not answer question 2?) If this type of nonrandomness exists, the results of the analyses can be severely affected. BMDPAM can use those methods available in BMDP8D. It also can replace the missing values with simple means among the data, means within a given category of the data, or values predicted from regression equations. Statistics on the reliability of the predictions are also given.

The handling of missing data is often a very critical step in a data analysis. If the amount of missing data is minimal, most methods given above can be used with similar results. However, as the amount of missing data increases, some type of regression analysis, such as is available with BMDP8D, usually proves to be superior.

In SAS, a decimal point is used to indicate a missing data point. Blanks or decimal points in variable format locations are interpreted as missing. Alternatively, the missing value specification can be defined by the user. Statements such as "If X = . THEN DO;" can be made for designating how missing values are to be handled. Internally, SAS uses a value of minus infinity for missing value data, causing these to be sorted as the smallest value. No calculations will be performed if any of the necessary values are missing. Those cases are usually eliminated (this is somewhat subroutine specific) and calculations are performed only for complete cases. No internal programs as such for estimating the values are available.

It must be remembered that there may be no answer to the question of how to represent the true analysis that would exist if all data were present. Filling in missing values, although often necessary, is only a means for allowing the cases to be included in further analyses and cannot guarantee their accuracy.

---

*Available from the BMDP address in Appendix V. *Missing Data and BMDP: Some Pragmatic Approaches*, Report #45 by James W. Frane.

APPENDIX IX
STANDARD SCORES AND
WEIGHTINGS

Both methods to be considered here are data modification techniques that actually change the values of the data used for the analyses. At first glance it may seem that they should not be considered simultaneously since standardization of scores is a technique that is strongly suggested for use, and weightings are used only with extreme care. However, both techniques do have their correct place in total data analysis problems.

Standard scores are also referred to as z-score transformations or scaled data. These are used to equalize the effect of each variable in the analysis. For instance, assume that two variables are being considered in a chemical analysis. Assume that the values for variable A range from 0% to 80%, and for variable B from 0 to 1000 parts per million. Raw data are fed into the computer program as the percent or parts per million analyses for the variables. Now consider the difference between a measurement of 8 and 9 for each variable. For both variables this reflects an absolute change of 1. However, in variable A, this is equivalent to 1.3% of the range (1/80) and in variable B to only 0.1% of the range (1/1000). The difference lies totally in the units in which the analysis is reported. If only raw data are used in the program, both changes of one unit will be considered equivalent when they really are not.

Some type of scaling is necessary to correct for this. One can think of scaling as the technique automatically used when making two-dimensional plots. If the variable to be plotted on the x axis has a range of 0 to 80, and that on the y axis a range of 0 to 1000, each division on the x axis may be made to represent one unit while that on the y axis represents ten units. This makes both axes approximately the same length while still representing different magnitudes of values.

The normal way to transform these variables in a way to give each equivalent importance in the data analysis problem is to generate a new variable from the old one that now has a mean of zero and a standard deviation of one. The values for each case then represent a number of standard deviation units that analysis is above or below the mean of the variable. This is then used in subsequent analyses instead of the raw data.

This procedure often utilizes the following equation:

$$z = \frac{x_i - \overline{x}}{s}$$

where z is the z-score, x_i is the raw data measurement for ith case, $\overline{x}$ is the mean value for that variable, and s is the standard deviation for that variable. This will be the value obtained from SPSS. For ARTHUR, the value is calculated similarly. However, the above result is then multiplied by the factor $(N\text{-}1)^{1/2}$ where N is the total number of degrees of freedom present in the problem.

Comparisons can be made between variables by means of Z-score transformations. Since each now has a similar distribution and value range, each becomes equally important in the data analyses. In some applications, z-score transformations are not necessary. In others, however, meaningless results are often obtained unless the transformation is made.

In ARTHUR, standard scores are calculated in subroutine SCALE. This output is then used for the input into the other statistical routines. In most of ARTHUR's procedures, multivariable techniques do utilize all of the variables simultaneously. No *a priori* prejudices towards certain variables are desired. Therefore, for most ARTHUR techniques, scaling should be used. This includes all supervised and unsupervised steps.

In SPSS, z-scores are available through CONDESCRIPTIVE. The z-scores form a file and can be written on punched onto cards. The file can be saved for subsequent analyses using a "SAVE FILE" command. In BMDP, standard scores can be found in BMDP1S which can also be saved for later use.

Another place to find the z-score transformation is in the options within the various techniques of BMDP and SPSS. The manuals will direct the user to these options. A rule of thumb is that they should be used when comparing variables. Examples are all supervised and unsupervised techniques including principal component analyses, and factor analyses. Since correlations are already scaled from -1 to 1, it is not necessary to use z-scores for these calculations.

In contrast, weightings are used to make variables of unequal importance. *A priori* prejudices are added to the importance of the variables. One of the times that such a process may be acceptable is during supervised learning techniques. In this process, the goal is to train the computer to recognize the accepted categories. Variables can then be weighted according to their relative importance in affecting a separation of classes. A method exists in ARTHUR called WEIGHT which will weigh samples this way. The variance is divided as interclass or intraclass for each variable. If these two are equal, a value of one is assigned for a weight. As the intraclass variance exceeds the interclass amount, a higher weight is assigned. If the variable affects an excellent separation, a value as high as ten can be obtained. However, normally the dynamic range of the weights is not great and this type of weight does not drastically change the features.

A Fisher weight can also be used. This method has a much larger dynamic range (up to 10^5). The means of a pair of categories are subtracted, and divided by the square of the intraclass variance. This technique is more sensitive to overlapping categories (those with similar means), but for separable data, results are similar to those obtained with variance weights.

Weightings should not be used with unsupervised learning techniques. Unsupervised steps are used to find natural groupings in the data. If no *a priori* knowledge is assumed for category structures in the data, one cannot know *a priori* what variables affect the group separation the most. Weightings are also inherently included in some of the mathematical techniques such as regression and factor analysis. In these cases, weights are not determined by the user, but rather by the statistics built into the mathematical manipulations themselves.

A different type of weighting is available in SPSS using a WEIGHT specification code. This can also be done in BMDP. In these methods, cases instead of variables are weighed. If *a priori* knowledge is known about the relative importance of various cases, this can be used. However, again, it should be utilized only with extreme care. A good rule of thumb is not to use weightings and to use standard scores unless the problem dictates differently.

SUBJECT INDEX

COMPUTER PROGRAM INDEX